The Social Life of Connectivity
in Africa

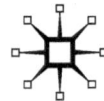

First published in 2012 by
PALGRAVE MACMILLAN®
in the United States—a division of St. Martin's Press LLC,
175 Fifth Avenue, New York, NY 10010.

Where this book is distributed in the UK, Europe and the rest of the World,
this is by Palgrave Macmillan, a division of Macmillan Publishers Limited,
registered in England, company number 785998, of Houndmills,
Basingstoke, Hampshire RG21 6XS.

Palgrave Macmillan is the global academic imprint of the above
companies and has companies and representatives throughout the world.

Palgrave® and Macmillan® are registered trademarks in the United
States, the United Kingdom, Europe and other countries.

ISBN: 978-1-137-27801-2

Parts of the Introduction are adapted from an earlier version that appeared
in the journal *Anthropologica* (de Bruijn and van Dijk, 2012) and are
reproduced by kind permission of the publisher.

Figures 10.1 and 10.2 © Rufus de Vries (www.rufusdevries.nl) are
reproduced by kind permission of the photographer.

Library of Congress Cataloging-in-Publication Data

The social life of connectivity in Africa / edited by Mirjam de
    Bruijn and Rijk van Dijk.
        p.   cm.
    ISBN 978-1-137-27801-2 (alk. paper)
    1. Communication—Social aspects—Africa.   2. Telecommunication—
    Social aspects—Africa.   3. Africa—Social conditions—1960–
    4. Social integration—Africa.   5. Globalization—Africa.
    6. Information society—Africa.   I. Bruijn, Mirjam de, 1962–
    II. Dijk, Rijk van, 1959–
    P92.A35.S63 2012
    302.2096—dc23   2012040654

A catalogue record of the book is available from the British Library.

Design by Integra Software Services

First edition: December 2012

10 9 8 7 6 5 4 3 2 1

Transferred to Digital Printing in 2013

The Social Life of Connectivity in Africa

Edited by Mirjam de Bruijn and Rijk van Dijk

Contents

Figures

Notes on Contributors

Astrid Bochow is an anthropologist currently based at the Max Planck Institute for Social Anthropology in Halle, where she is conducting research on issues of involuntary childlessness in Botswana. Her previous research work focused on youth, courtship patterns, sexuality, and religion for which she did extensive fieldwork in Ghana. Her PhD has been published recently by LIT-Verlag, Hamburg, entitled *Intimität und Sexualität vor der Ehe. Gespräche über Ungesagtes in Kumasi und Endwa, Ghana* (2010). She has furthermore published extensively on premarital relations of youth in Ghana in the context of Valentine's Day celebrations and the rise of Pentecostalism in the country. An important publication derived from this work is "Valentine's day in Ghana: Youth, sex and fear between the generations." In: Alber, Erdmute, Sjaak van der Geest & Susan R. Whyte (eds.) (2008) *Generations in Africa*, Hamburg: LIT, pp. 418–429.

Angélique Bos has studied Work and Organizational Psychology at the University of Groningen in the Netherlands, where she obtained her BA degree. She then moved to the VU University in Amsterdam, where she obtained an M.Sc. degree in Culture, Organization, and Management. During her M.Sc. study, she spent four months in South Africa conducting research on public-private partnerships in land reform in Limpopo Province. She currently holds a position as Management Trainee at ORMIT, a training program for the development of management competencies. She has an article forthcoming in the land reform newsletter *Umhlaba Wethu* entitled "Government, land reform beneficiaries and private organizations; joining hands in the struggle for land restitution?"

Inge Brinkman is a historian and Africanist (PhD, Leiden, 1996). She has been engaged in various projects on Sociocultural History at the African Studies Centre, including the history of development work, the war in Angola, and communication technologies in Africa. At present she is one of the coordinators of the research program "Mobile Africa revisited," which focuses on the relations between mobility, communication technologies, and social hierarchies in various so-called marginal regions in Africa. Among her recent publications are " 'Communicating Africa.' Researching Mobile Kin Communities, Communication Technologies, and Social Transformation in Angola and Cameroon", *Autrepart*, 57/58 (2011) pp. 41–58 (coauthor: Mirjam de Bruijn), and "UPA Pamphlets and Politics in Northern Angola: Changing Concerns, Changing Messages, Around 1961," in: Joseph C. Miller, Philip J. Havik and David Birmingham (eds.) (2011), *A Scholar for All Seasons: Jill Dias, 1944–2008, Special Double Issue, Portuguese Studies Review*, 19, 1/2: 293–310.

Ben Cousins is Professor in the School of Government at the University of the Western Cape where he holds a DST/NRF Chair in Poverty, Land and Agrarian Studies. He has worked in agricultural training and extension in Swaziland and Zimbabwe. His research over the past decade has focused on the key themes of production, property, and power, and their interconnections in the context of land and agrarian reform in Southern Africa. This research is strategic and use-oriented, in this instance by policymakers and civil society groups concerned to reduce poverty and inequality through redistributing assets, securing rights, and democratizing decision making in rural areas. His publications include an edited volume, coedited with Aninka Claassens (2008) entitled *Land, power and custom: Controversies generated by South Africa's communal land rights act*, published by Ohio University Press.

Mirjam de Bruijn is a senior researcher at the African Studies Center in Leiden, The Netherlands, and Professor of African Studies (Contemporary History and Anthropology of Africa) at Leiden University. She has been conducting research in various countries in West and Central Africa including Cameroon, Mali, Chad. Her main interests are Social Anthropology, History, African Studies, (Post-)Conflict, Mobility/nomadism, Processes of inclusion and exclusion, ICT (information and communication technology). She was awarded several research grants, the latest being a Vici grant (by the Dutch Foundation of Scientific Research) for a five-year research program on ICT (social media and mobile telephone) and conflict in Africa that started in September 2012. She was coeditor of the first book on mobile telephony in Africa (with Francis Nyamnjoh and Inge Brinkman) *Mobile phones: The new talking drums of everyday Africa* (Langaa Publishers, 2009). One of her recent articles (with F. Nyamnjoh and T. Angwafo) (2010) is "Mobile interconnections: reinterpreting distance, relating and difference in the Cameroonian Grassfields" *Journal of African Media Studies*, 2(3): 267–285.

Danielle de Lame, with a background as a sociologist, presented her PhD in social anthropology at the Free University in Amsterdam. She is the head of a research unit at the Royal Museum for Central Africa (Tervuren, Belgium). She is the coauthor of the Rwandan national bibliography and a specialist of Rwanda where she did her main fieldwork. More generally, she studies the dynamics of social change and has published extensively on Rwanda, with other publications on Senegal, Congo, and Kenya; a forthcoming book under her direction takes aspects of South African "urban cultures" in consideration as well.

Jan-Bart Gewald is a historian specialized in the social history of Africa. His research has ranged from the ramifications of genocide in Rwanda and Namibia, through to the sociocultural parameters of transdesert trade in Africa. In addition, he has conducted research on pan-Africanism in Ghana, spirit possession in the Republic of Niger, Dutch development cooperation, Africa in the context of globalization, and social history in Eritrea. Furthermore, he has a particular interest in archaeology, and has participated in archaeological research in southern Africa. He is currently working on the social history of the automobile in Zambia. A recent publication is J. B. Gewald, Sabine Luning & K. van Walraven (eds.) (2009) *The speed of change: Motor vehicles and people in Africa, 1890–2000*, Leiden (etc.): Brill, Afrika-Studiecentrum series; vol. 13.

Mayke Kaag is a researcher at the African Studies Centre. She holds a PhD in social anthropology (2001) and has a strong interest in processes of change and continuity in West and Central West Africa. Among other things, she conducted research into land use and social change, international migration, and transnational Islamic charities in Africa. Among her recent publications are "Aid, *Umma* and Politics: Transnational Islamic NGOs in Chad." In: R. Otayek & B. Soares (eds.) (2007) *Muslim politics in Africa*, New York: Palgrave Macmillan, pp. 85–102, and "Transnational Islamic NGOs in Chad: Islamic Solidarity in the Age of Neoliberalism", *Africa Today* 54(3), pp. 3–18, 2008.

Walter Gam Nkwi is a social historian who graduated from the University of Buea, Cameroon, with a Masters degree in African History. He holds a PhD from the University of Leiden, The Netherlands (2011). His thesis is entitled "*Kfaang*: An emic history of inter-relations between Information Communication Technology, Mobility and Identity dynamics in Kom (Cameroon) 1928–1998." He has published widely in peer review journals and his latest publication is *Voicing the voiceless: Contributions to closing gaps in Cameroon* *history, 1958–2009* (Cameroon: Langaa Research & Publishing CIG, 2010).

Lubabalo Ntsholo holds an M.Phil. in Land and Agrarian Studies from the University of the Western Cape. He is a Programme Developer at the Succulent Karoo Ecosystem Programme (SKEP) of the South African National Biodiversity Institute. His main research interests are land reform, rural development, the political ecology of natural resource management, and the interface between biodiversity conservation and livelihoods development.

Francis B. Nyamnjoh is Professor of Anthropology at the University of Cape Town, South Africa. He has taught sociology, anthropology, and communication studies at universities in Cameroon and Botswana, was head publications at CODESRIA, and has researched and written extensively on Cameroon and Botswana, where he was awarded the "Senior Arts Researcher of the Year" prize for 2003. Prof. Nyamnjoh has published widely on globalization, citizenship, media, and the politics of identity in Africa. His book *Africa's media, democracy and the politics of belonging* (Zed Books, 2005) is a major contribution to the study of media in Africa. He has also published seven ethnographic novels.

Koblowe Obono teaches in the Mass Communication Department of Covenant University, Nigeria, after several years of academic practice at Bowen University and the Ibadan Polytechnic. Her disciplinary backgrounds in Language, Communication, and Sociology account for the distinctive interdisciplinary perspective in her work on health communication, language use, media analysis, gender, and its emphasis on the layers of human behavior. She is a Fellow of the African Humanities Programme and Principal Investigator of the study *Sculptures, Sounds, Symbols and Scripts of Sexuality*, sponsored by the American Council of Learned Societies.

Oka Obono is Professor of Sociology at the University of Ibadan, Nigeria. His research interests span more than 20 years of work in the African field and emphasize the fluid and fragmented quality of life—the "mismanagement of existence"—in modern Africa. His weekly newspaper column, *Predicaments*, examines the impacts of a postcolonial state and activist civil society on everyday life. He is editor of the multicountry collection of empirical studies, *Tapestry of Human Sexuality in Africa* (2010), an intricate examination of understudied elements of sexuality on the continent, which include links between distress and intimacy; the persistence of genital cutting and the motivation for mutilation; crises of shelter and identity; and online connections of religious communities around issues of sexuality. He is Director of the Millennium Advancement Initiative (MAI), a regional network of African and Africanist researchers, and is best known for his work in research methodology, ethnography, policy analysis, and literature.

Neil Parsons is emeritus Professor of History at the University of Botswana. His books include *The roots of rural poverty in Central and Southern Africa* (Heinemann Educational & University of California Press, 1977), *A new history of Southern Africa* (Macmillan Education, 1982 & 1993), *Seretse Khama 1921–1980* (Macmillan Boleswa, 1995), *King Khama, Emperor Joe, and the great white queen: Victorian Britain through African eyes* (University of Chicago Press, 1998), and *Clicko the wild dancing bushman* (Jacana, 2009 & University of Chicago Press, 2010).

Marja Spierenburg is Associate Professor at the VU University Amsterdam, and the University of Stellenbosch in South Africa. She is engaged in several research projects focusing on the role of the private sector—both for-profit and nonprofit—in nature conservation and land reforms in Southern Africa. Together with Shirley Brooks, University of the Free State, she coordinates a research project funded by the Netherlands Organisation for Scientific Research (NWO) that addresses the impacts of conversions from "conventional farming" to game ranching on the livelihoods of farm workers and (former) labor tenants in KwaZulu-Natal and the Eastern Cape, South Africa. Her publications include "Conservative philanthropists, royalty and business elites in nature conservation in southern Africa," coauthored with Harry Wels, published in *Antipode* (2010), and a monograph entitled *Strangers, Spirits and Land Reforms; Conflicts about Land in Dande, northern Zimbabwe* (Brill Academic Publishers, 2004).

Walter E. A. van Beek is an anthropologist at the African Studies Centre in Leiden and Professor of Religious Anthropology at Tilburg University. He has done extensive research on two West African groups, the Kapsiki/Higi of North Cameroon and Northeastern Nigeria, and the Dogon of central Mali, with religion, ecology, and tourism as the main themes. His current research links local religions to the wider dynamics of religion in the world. Recent publications include *The Dancing Dead; Ritual and Religion among the Kapsiki/Higi of North Cameroon and Northeastern Nigeria* (New York, Oxford University Press 2012), and *African Hosts and Their Guests; Dynamics of Cultural Tourism*, coedited with Annette Schmidt (James Currey, Oxford, 2012).

Linda van de Kamp (PhD VU University Amsterdam, 2011) is a cultural anthropologist and a postdoctoral researcher at the department of Cultural Studies, Faculty of Humanities, Tilburg University, The Netherlands. Her research activities concentrate on religion and ritual, South-South transnational connections, gender and reproductive issues, cities, violence, and insecurity. Linda has done in-depth research on the emergence of Brazilian Pentecostalism in Mozambique. Recent publications include "Converting the Spirit Spouse: The Violent Transformation of the Pentecostal Female Body in Maputo, Mozambique" (*Ethnos* 76(4): 510–533, 2011) and "Pentecostals Moving South-South: Ghanaian and Brazilian Transnationalism in Southern Africa" (Brill, 2010, with Rijk van Dijk).

Rijk van Dijk is an anthropologist working at the African Studies Centre, Leiden. He has done extensive research and published on the rise of Pentecostal movements in urban areas of Malawi, Ghana, and Botswana. His current research deals with the religious, in particular Pentecostal, engagements with the domains of sexuality and HIV/AIDS in Botswana. A recently published article entitled "Gloves in times of AIDS: Pentecostalism, hair and social distancing in Botswana" (In: F. Becker & P. W. Geissler (eds.) (2009), *Aids and religious practice in Africa*, Leiden/Boston: Brill, Studies on Religion in Africa,) deals with insights gained from this ongoing research. In addition, he is the editor-in-chief of the newly established journal *African Diaspora. A Journal of Transnational Africa in a Global World*, which is published by Brill, Leiden, as of 2008.

Acknowledgments

This book marks the end of a collaborative project between colleagues who worked together in the Connections and Transformations research group at the African Studies Centre in Leiden for four years (2008–2012). Without the cooperation and collegiality of this group, this book would not have been published. The contributions to our research collaborative were numerous. Thoughtful exchanges developed with a number of people who became our advisors in the writing process: Erdmute Alber, Kurt Beck, Georg Klute, Annemarie Mol, Dieter Neubert, Francis Nyamnjoh, Oka Obono, Chris Reina, and Ria Reis. Their contributions to the various (inter)national workshops, e-mail exchanges and monthly meetings that formed part of the bringing to maturity of many of the ideas presented in this book were invaluable. Not all our colleagues from the research group were able to write an article for this volume but their contributions to our discussions have certainly found their way into the various chapters. We therefore express our deep gratitude to Daniela Merolla, Laurens Nijzink, Samuel Ntewusu, Lotte Pelckmans, and Doreen Setume. We would also like to thank Ann Reeves for her language editing and Karin van Bemmel, Inge Butter, Kim van Drie, and Roos Keja for helping with the production of the book and for finally making this project happen. Our special gratitude goes to the Bayreuth International Graduate School of African Studies (BIGSAS) that sponsored two workshops, and to the African Studies Centre that guaranteed our basic research funding. We especially remember the warm welcome the group received from Mr. and Mrs. de Bruijn who invited us to hold one of our writing sessions in the intimacy of their home. The program has benefited from grants that were awarded to members of the group by NWO (Netherlands Organisation for Scientific Research). Our hope is that our connectivity will produce other interesting products in the future! A big thank you to all of you!

Introduction

Connectivity and the Postglobal Moment: (Dis)connections and Social Change in Africa

Mirjam de Bruijn and Rijk van Dijk

Coming from the district of Sor, on the continent, you can access the island of Saint-Louis only by one link, the "Pont Faidherbe." This metallic structure is 507 meters long and 10.5 meters wide. The bridge of Saint-Louis consists of 7 arched spans with the second to last having been conceived to pivot around a fixed axis to let ships pass through. Two wooden sidewalks border its central pass-way. Over one hundred years old, this bridge was first inaugurated on October 19, 1897.[1]

Introduction

This bridge (Figures I.1 and I.2) inspired me when I (Mirjam) was in Saint-Louis in 2009. It connects Saint-Louis to its colonial past. Louis Faidherbe was the Governor of French West Africa in the nineteenth century and is perceived as one of the "creators" of this relationship. The colonial regime has clearly had an impact on present-day Senegal and, in a way, the bridge represents a connection between the past and the present. The bridge is memory and a memory of the technologies that the French brought. It is a memory of colonial times.

The bridge also connects Saint-Louis with the mainland and 24 hours a day people cross the bridge on foot and by car to go to work or to go home, to visit friends, or just to cross it. The bridge was a technological innovation that made the passage easy and replaced boats. When this bridge

Parts of this introduction are derived from an earlier publication in *Anthropologica* (de Bruijn & van Dijk 2012), which we hereby gratefully acknowledge.

Figure I.1   Faidherbe Bridge, Saint-Louis, Senegal, © Mirjam de Bruijn.

was being constructed it must have generated amazement and emotions that made people believe in progress, in "development."

I stood and observed life on the bridge for a while: people stopped to meet and talk; sometimes it was clearly for business, at other times it was social talk. As a crossing point for many people, the bridge is also the ideal spot for a beggar to dwell. Of course the bridge was built to provide transport from one place to the other, by car or on foot. Walking has styles, as do cars. Taxis, big trucks, and the many personal cars that cross the bridge all give signals to the observer about the well-being of the people and the trade and public transport of this part of Senegal.

The bridge is no longer cutting-edge technology but has become part of the social and economic life of Saint-Louis. It has become more than just a bridge connecting two areas; it has become social life itself in the actions I observed and in its memory of the past.

Understanding the bridge connection as a social, economic, political, and historical process helps one to understand the changes that have taken place in Saint-Louis. Without this bridge, the city would not have developed in the way it has and its inhabitants would not have become who they are. This "becoming" is not so much related to individual agency and actions but to sharing, communication, to being connected, as is symbolized by the bridge. The activity of relating to these connections, or to this communication,

Figure I.2 A beggar on the Faidherbe Bridge, © Mirjam de Bruijn.

transforms the connection, both in its materiality and in its social and economic effects.

An important point of departure in understanding the bridge and the changes in Saint-Louis is not the people living on the two sides of it and who use it to cross the river but the connection itself and how it is socially, politically, and economically appropriated, as well as how it seems to have a life of its own.

The metaphor of the bridge helps us to understand that linkages between people are important as the connections themselves appear to have transformative powers. Invoking Appadurai's the social life of things (see Appadurai 1986, see also van Binsbergen & Geschiere 2005), these connections have a social life. And understanding how this is formed may help us understand how social transformations in society and connections are related.

Studies of today's globalizing world have concentrated on analyzing the network and information society (see de Lame, this volume), with questions about the information society, the network society, and its social meaning focusing in particular on understanding the "people" involved in these networks. The idea of networking is central to our notion of the importance of connections, yet in our search of processes of social transformation from a connections perspective, we move explicitly to the significance of linking. In processes of linking and making new connections and decisions, we can see new social constellations coming into being and social transformations taking place. The connection has a bearing on the people concerned, which means that by moving away from network theory—in which agency is placed fully in the people who are being connected—it is the connection itself that plays a role. We need to allow for the possibility that it is the connection that enables a new constellation to emerge, and it is not the "dots" in the network that are being connected.

Our interest in the way people appropriate connections stems from our work on agency (de Bruijn *et al.* 2007), where the actor and his/her relating are central. In agency research as well as in network research, the focus is on the actor and what is being connected. But it is the nature of the connection and the way people come to embrace connections that may help us understand social transformations. It is important to note that the nature of the connection is often overlooked in anthropological and sociological theories. A general model to understand social transformation from the perspective of the social life of connections and a model to describe moments in the history of such a process have been lacking. This is remarkable given the new age of technology, the level of success of so many new technologies of connections, and their rapid spread around the globe that has allowed access to so many groups to connections. We believe that connections have become important in the social fabric in Africa and beyond at a juncture that we would label "postglobality." With this term we indicate that connections have played, and are still playing, a crucial role in how the global became ordinary, also in Africa. In the following paragraphs we will elucidate this process.

Connections in the Age of Postglobality

Figure I.3 shows a social situation that inspired us in our thinking about the centrality and agency of connections in Africa.

This medicine man, who is connected to the spirit world by phone, illustrates something beyond globalization, something we term "postglobalization." The connection has a power of its own that makes possible the transformative process of healing for which the medicine man is responsible.

Figure I.3   Sing'anga in Blantyre, Malawi, © Rijk van Dijk.

The idea that globalization is a new process is negated in this picture: African realities have become globalized through the appropriation of technologies. The intensification of exchange, travel, mobility, and flows of objects and ideas that all came under the term "globalization" can no longer be considered a new phenomenon in Africa. The conceptual study of globalization, as it has developed since the 1990s, has been pointing at a range of remarkable "intensifications," that is, the enormous increase in transnational travel, mobility, and connections, an unprecedented rise in information and communication technologies of which the most spectacular is the mobile phone, the significance of new patterns of consumption, and the flow of global images and ideas. The local cannot be considered without the global in Africa today and local realities are being shaped and reshaped in view of global connections.

These intensifications have given way in both rural and urban areas or are being produced by an abundance of connections. In many instances, groups of different socioeconomic status, gender, and age have become intensely connected. Villagers are connected with migrants overseas and through these diverging connections money, objects, and ideas are circulating easily and rapidly. In addition to the way African settings were connected to an outside world in the past, a new and highly vibrant tapestry of connections through the media, modern means of travel, and communication and a wide-ranging exchange of goods and styles has come into being.

This process of realization of the abundance of connections in Africa is reinforced by ongoing developments in infrastructure. The building of roads, railway lines, bridges, the process of electrification, increased coverage of media and communication technologies, and intensified activity by international organizations, NGOs and faith-based organizations (FBOs) are all part of this expansion. In areas of conflict and violence such expansion has clearly been brought to a standstill but even in war-torn areas the arrival of the mobile phone and the required infrastructure has not gone unnoticed (see Brinkman & de Bruijn, this volume).

This combined process of intensification of connections and the expansion of infrastructure also forms the basis of the notion of a postglobal society in Africa where the creation of a globalized society has been realized and the multitude of connections is a fact and has become part of the experience of most Africans. Connections are no longer a scarce resource.

When speaking of postglobalization, we have to understand what difference the nature and quality of the linkage itself makes in terms of social formations. The contributions in this volume demonstrate how globalization implies the emergence of an expanding awareness of what linking does and means in a modern world. While variations of linking have increased enormously, the need for people to consider what each of these forms of linking can and cannot do or mean has become of much greater importance.

The following sections develop questions that relate to linkage analysis as it is necessary to understand the connection, the appropriation of the connection, and the social transformation or nontransformation it entails. We focus on the connections and how to understand these once they have been internalized and become part of the social fabric. After a comparative analysis of space and time, networks and social capital are discussed, more specifically ideas of bonding and bridging (Portes 2000; Putnam 2000) and those of weak and strong ties as put forward in network theory (cf. Granovetter 1973). And the relationship is explored between connections and wealth (Guyer 1995). When discussing the concept of connections, the inherent power relations that all connections represent is ever present (cf. Ferguson 2006; Castells 2004). Connections are never a neutral phenomenon but in their appropriation by people, groups, and institutions form part of power hierarchies in ways that are informed historically.

Questioning Connections

In a research program on the changing forms of connections in a postglobal society and how changes occur in social relating and ideologies, the connection itself should first be deciphered. "Connection" is a broad concept that can mean many things but what is important is the significance of linking: connections indicate a bridge between objects, between things, and between humans. This introduction has mentioned the subjectivity of the bridge in Saint-Louis as an important field of exploration beyond its materiality of how it was made and what kind of connections it offers. Who uses the bridge, how is it interpreted and integrated in the social life of people?

Connections—or bridging—can be described as objects; they are always changing, flexible, and varied but objects can be appropriated by a society, an individual, or in dreams. Part of knowing the connection is to understand what brings about the connection. Connections should not only be related to concrete changes in technologies. The colonial period brought bridges, roads, and cars and telephone lines to Africa and certainly influenced "connecting" in the process but it also brought institutions like the Church, schools, and clinics that also created new forms of connecting. The materiality of connections is therefore important in their subjective interpretation. The immaterial nature of certain connections and their transformative efficacy, such as connections established through religious and political ideologies, power, and sensory regimes, have been little explored. Material connections appear easier to understand.

Connections become pivotal in the way social relations are shaped or refigured. We are only just beginning to understand the significance of what it means for societies when new connections are established, old connections change or disappear, or certain things and places become disconnected. Research into connections is focusing on the processes by which connections become "useful." Such connections are seen as opportunities or are given meaning by the people involved or those who are *not* involved (the disconnected). This is where connections turn into social or political relations or networks. They can be studied by using well-established analytical tools such as network analysis, pathways, social relations, kinship, or hierarchies, provided they continue to be relevant. These models, however, have never really addressed the "how," the "what," and the "why" of newly created connections, and thus the social transformations they bring about (Strathern 1987, 2004; Green 2002). Are these connections properly understood? Why are people, communities, and institutions searching for new connections and how are these related to existing forms of agency and social change?

By questioning connections, we want to emphasize that connections and relations are not the same. Feldman recently argued in a similar vein:

> Why the preference for connections over relations and what are its consequences? The question is best answered by reminding ourselves of the difference between them even if they are mixed together. The difference is primarily that connections involve direct, immediate contact between people while relations involve indirect, mediated contact. People historically have been linked through direct social connections, characterized by their tangibility, corporeality, and locality.
>
> (Feldman 2011: 397)

Connections make different relations possible. When trying to grasp the meaning of new options and possibilities for connecting, the most obvious process to study is "relating." What we take here is the sociological meaning of the term: how do people relate and thus form a society, community, or a social network through connections? If new possibilities for connecting are introduced in a social space, how is the process of relating then shaped? People rely on their long-established ways of relating but will also employ new ways of relating and invent forms of relationships that may not have existed before.

People have reasons for connecting and disconnecting and do so in meaningful ways that are informed by the cultural and social repertoires in the specific societies that are relevant to the connections. These processes of change cannot be understood without analyzing people's agency, which focuses on individual and collective decisions, interpretations, reactions to and reflections on changes in their environment, or without understanding the ways in which these interactions shape their social world.[2] Individual and collective agency shows how people and institutions work together, react to, and impinge on each other by establishing various types of connections. In the process, connections become part of the new social hierarchies, different (labor) relations, various forms of cooperation, and new forms of mobility. When questioning connections, we need to explore their apparent agency. Connections seem to represent a resource that sets things in motion, enables transformations, and builds new power relations.

Connections as a Resource

Connections have become so readily available that there is no longer any reason to consider them a scarce resource. All these connections do indeed form resources to a varying degree for people in society, for example, the mobile phone offers access to markets and important contacts, and a bridge provides access to hinterlands or the Internet might provide access to job opportunities. Since the connections are there and connectivity cannot be perceived as a scarce resource, does it mean that studying these connections has become less valorized for the people in African societies?

We argue that with the increasing availability of connections, the global and the presence of global identities, issues, objects, ideas, and images have become ordinary in African societies of today. By calling this a phase of "postglobalization" we notice how this has changed the relative value of connections. As globalization is a theory and a process of intensification (of circulation, travel, and mobility), some connections are becoming more valorized than others precisely in the way some have become ordinary, which they were not before. People have begun to take it for granted that they are prepared to spend more on some connections that have become available through new technologies than on others, as these connections position them (better) in global flows. Or, on the contrary, they may prefer one certain connection more than another because it has become less expensive or because it has increasingly become more common to use, and therefore a reduction in cost has followed. A question that people can be faced with, such as "will I connect with a bus to that other town or shall I call?," indicates a situation in which, first of all, an increasing number of connections have been made available and, second, an economy of scale and "commonality" of the connections involved and a level of the intensity and relative value of the connection in social terms have occurred.

While both of these means of connecting in this example—the bus and the phone—have been made possible by the global spread of modern means of transport and communication, the modes of connecting and linking are differently valorized. In many historical cases, connections were of great value, as was the case of the coastal-bound railways in Africa that transported agricultural produce or natural resources from inland areas to ports for shipment to Europe. The colonial exploration and exploitation of Africa's riches was made possible by investments in connections so that the accumulation of wealth required for the connection would generate new forms of accumulation elsewhere. The arrival of the motor car in Africa also connected people in new and innovative ways but at the same time was an instrument that intrinsically represented enormous wealth.

While the formation, construction, and exploitation of infrastructure in African societies in colonial and postcolonial times have been studied in depth, the point about connectivity in this respect is not so much its economic analysis. While acknowledging that connections may be the result and product of wealth and value, and in turn can generate more accumulation, what the research program has been exploring is their social construction, representation, and embeddedness in social relations. Connections and their related infrastructure may be of importance in other domains of value than only an objective economic one. The wealth in connections that are perceived emically should be studied in their own right. These receptions, sentiments, and values of connections are never given in advance but are likely to be structured along the relations, the scope, and the nature of those that they foster. The wealth of a bridge, as in the example, may go beyond its obvious economic value as it may foster relationships, kinship, rituals, and the movement of ideas, and its value can thus be explored in terms of culture, sentiment, affection, support, and reciprocity.

Little research has yet been done on how people in certain African communities perceive the wealth of connections. In anthropology, work has been done on so-called road mythologies whereby specific concerns of a symbolic and religious nature have been explored concerning the dangers of the road, its occult economy, and its relevance for the passage of evil powers (Masquelier 1992, 2002). This literature has often emphasized the exotic. The only way some African societies appear to understand and represent the products of modernity is by incorporating them in the domain of witchcraft and other phantasms. However, this is not the only way people's concerns with the wealth of connections should be discussed. Much can be taken from Guyer's (1995) perspective of the wealth in people. The way Central African societies relate ideas to wealth and prosperity is by considering which futures are being opened up and which avenues are being guaranteed for agricultural produce, reproduction, and labor in times to come by safeguarding a certain level of population growth. Connectivity is the constant negotiation of a level of wealth in connections. In this perspective, roads are not the concern of people as being primarily the passage for witches but a concern that puts the extent to which a range of economic, social, or cultural relationships can be maintained as central, and the road is a crucial instrument. These concerns can likewise be studied for bridges, mobile phones, and schools.

The wealth in connections is in this view directly related to how far social life is made possible, furthered, or terminated. It is also from this perspective that disconnections become a matter for emic representation too. These may be the cause of anxieties, a matter for strategic negotiation, or one of great interest, just like connections. Hence the concern for a wealth of connections does not exclude the possibility of the importance of disconnections. On the contrary, disconnections must be taken as a possibility and they put into relief the "why" and "how" of local concerns with connectivity.

While connections and wealth have been studied from a historical perspective as much of the colonial concern with extracting riches from Africa was dependent on how connections supported capitalist concerns, a contemporary and anthropological perspective of the wealth in connections is yet to be developed. The changing way in which connectivity as a social concern can be studied over time to map important social and structural changes is still in its infancy. These lacunae are noteworthy as it can hardly be assumed that connectivity was the concern of colonial and capitalist regimes and powers alone.

Connections Embedded in Power Relations

Any exploration of a postglobal world in Africa where people negotiate their engagement with connections has to be placed in the context of the common problems and predicaments of many African societies. Increasing differentiation in wealth, access to resources, poverty, and population growth set the conditions for connections. Increased connectivity also sharpens these differences and contestations. Connections may have the capacity to render positions of deprivation and inequality more visible. While connections in a sense "bring globalization home," they have an agency in terms of aggravating inequalities. Valorization also has a flip side, as is predicted by Ferguson (2006) as "global shadows" or by Castells *et al.* (2007) as the "Fourth World."

It is important for the contemporary study of connectivity to come to an understanding of how and under what conditions wealth in connections can become the subject of power relations at a local level. Connections and concerns about a wealth of connections are never neutral simply because they are resources, the subject of decision-making processes, and are influenced by particular ideologies and can be related to specific authority relations. Development efforts and interventions fostered by national governments in Africa concern the improvement of infrastructure, such as the building of roads, bridges, and communication lines. Increasingly, people's concerns with a wealth of connections have become enveloped in developmentalist rhetoric that usually state that it is to the greater benefit of all if infrastructural development takes place. While there is therefore a genuine interest in connectivities of various kinds, there is usually very little sensibility toward the emic understanding of what a wealth of connections may mean and how it should be fostered. It is often assumed, yet not proven, that the best way to improve and support a wealth of connections involves infrastructural development. We notice that this can even create a state monopoly in connectivity. The ideology of development may mask the detrimental effect of such a state monopoly on the economy, politics, human rights, and/or resources whereby connectivities can also be seen to play a role in the diffusion of the ideology, in the marginalization of dissenting voices, and in shifts in the balance of power toward the interests of particular groups in society.

Connections are never neutral regarding power dimensions. While mobile-phone connections appear to have engendered an almost democratic free-for-all availability that would enhance a wealth of connections, there is a veiling taking place of underlying power relations of control, economy, and authority that can be seen in the politics that surround the relationship between connections and wealth. While democratic in outlook, such as the general use of the mobile phone, greater disparities, inequalities, and uneven power sharing may result, damaging local notions of the wealth in connections. Disconnections for some people can result as money may be spent differently (on airtime instead of on food or transport for instance) or access to resources shared in other ways.

We propose studying the social and cultural appropriation of connections as a process of social transformation and as a political process embedded in existing social and political hierarchies. Being connected or disconnected thus becomes part of the process of the appropriation of connections.

Connections and Compartmentalization

An important element in the production of these political and other contestations and contradictions relates to how connections lead to new compartmentalizations of social reality. New connections may crosscut existing divisions in geographical and in more social, immaterial, and metaphysical terms. As Ferguson (2006) demonstrates, people and social groups or communities do not participate in the postglobal world in a similar way. Some are establishing connections easily, while others are marginalized in new and unprecedented ways. New compartmentalizations and perspectives of social order and society are emerging. If a person living in an African village connects with a brother in Europe by mobile phone or on the Internet, this connection produces a different sense of community and social order, much in the same way as the Internet does not provide connectivity with one's next-door neighbor. Partaking in diasporic connections may, therefore, demarcate a certain transnational social community. Glick Schiller (2005) termed this a transnational social field, which may make it a different compartment of social reality compared, for example, to a locally existing association. The impact of the connection, such as the Internet, on social life can be seen to relate to the emergence of contesting social compartments and to strengthening the notion of the social body, not as an organic whole but as segments.

At a theoretical level, there is the hypothesis that with connections comes an increase or change in the nature of social compartmentalization. Is compartmentalization a resource that connections bring? The following figures demonstrate this process, which is more easily understood from a geographical perspective. Places and spaces are increasingly being distinguished from one another when connections increase.

The notion of compartmentalization seems to emphasize connections as a material "thing" (a road, a bridge, a railway). There are two issues here. One is that connecting things are in no way equal to each other and big differences may exist in terms of cost, structure, and power. Does the nature of the connection reflect or even dictate the nature of the compartment? In the study on social compartmentalization, we need to consider the extent of the nature of new connections and how they influence the rise of new social compartments. In other words, new connections can lead to new social categories. And what happens if connections are not material? How can connections and their spaces be understood if they are not material and geographical but social and immaterial?

This reasoning links the discussion about connections to that of social capital. In a way, connections can be perceived as social capital as well (see Obono & Obono this volume). Social capital is seen to be related to functions of bridging and bonding across social compartments of any kind. A literature has emerged that explores the function of social capital and how it produces either a bridging or a bonding. Social capital is capable of uniting different sets of people and their interests (i.e., bridging) or producing exclusionary modes of social formation (i.e., bonding) that strengthen the sense of community but that may also lead to a greater juxtaposition of different groups in society. The literature that intends to judge societies accordingly, for example, Portes (2000) and Putnam (2000), can be seen as being highly normative in this regard, perceiving the positive or negative effects of social capital for the democratic content of Western societies but failing to answer basic questions about bridging or bonding that would be considered positive or negative.[3] While connections may lead to more compartments of social life, materiality, and social functioning, the concept of social capital appears diametrically opposed as it emphasizes the ways compartments, distinctions, and differences within society are being overcome.

The conceptual concern with the social construction of connections must position itself regarding the ways in which sociological analysis perceives the importance of bridging and bonding. Bridging and bonding presuppose access to knowledge of a domain that gives content to the connection at both ends (Hannerz 1992, 1996). Connections bridge distances, places, and people and their communities, and those such as the telephone, the car, or the church may bind people together in certain or new frameworks of identity. However, connections are not the same as social capital, nor is it possible to assume a normative standpoint on the quality of connections in ethically defined positive or negative terms. At a deeper level, however, connections allow us to explore the question of social capital in a different way. Instead of juxtaposing the functionality of connections as either bridging or bonding as mutually exclusive outcomes of a connection, a much more relevant issue is whether the bridging that connections make possible leads to particular or new forms of bonding. While the work of Portes (2000) and Putnam (2000) questions whether social capital fosters differences between people (either by bridging or by bonding in view of differences), the relevance of the question of connections and their social representation is how bridging is related to bonding in a processual manner. What type of bridging leads to weaker or stronger bonding? Which bonding requires the specific forms of bridging that connections provide? If a bridge connects two sides of a river, new bonds of an economic, social, or cultural nature may result but this bonding will not necessarily be stronger than the connection that a mobile telephone can provide.

Connections thus appear as a precondition for social capital, making it possible to consider whether a particular connection says something about the nature of a specific form of social capital that results from it or that is fostered by it. A mobile telephone is likely to have a different effect on the creation of social trust, confidence and reciprocity than a bridge, but this may be the result of the different principles and dimensions of bonding that the bridging provides.

This is highly relevant in the African setting as new connections can have huge repercussions on bridging and bonding, more so than would be expected in the West due to people's access to resources. The traditional doctor in Figure I.3 is connecting to something through his magical telephone; yet what is being connected to is largely immaterial. Underneath the spotted furs hanging from the wall and to which the wire of beads is connected live his ancestral spirits. The connection indeed seems to compartmentalize (i.e., between a visible and invisible world, between the world of praxis and spirituality, between the world of social relations and that of supernatural forces), perhaps in unexpected ways. So while connections may combine a material and an immaterial world, making compartmentalization a process at different layers at the same time, the bridging and the bonding that occur become multilayered.

One of the most significant immaterial connections can be found in ritual. Why can ritual be called an immaterial connection? One reason is that rituals usually have material dimensions and bring about a compartmentalization of social reality. Specific spaces are designated for ritual and it seems to carve out social space in definite ways. The more ritual there is, the more compartments there are! Does this reflect on the nature of the compartments? Lefebvre (1991) presents relevant ideas about the social production of space, as does Appadurai (1995) on the social production of locality. Ritual appears to compartmentalize social spaces for identity formation, informing questions about who belongs and who does not. Increased rituals therefore mean a mounting complexity of social spaces (see chapters of van Beek and Nyamnjoh, this volume).

The connections perspective leads to an increased understanding of the multilayered nature of concepts such as social capital, class formation, and processes of social positioning. The linkages that form part of the relations of trust, reciprocity, and associational life that are studied under the concept of social capital can be seen to be active at different levels at the same time. Also, the introduction of new moralities and social bonds that become activated in unexpected ways or sudden discontinuities and disempowerments can be understood from questions about what happens if connections cannot be mobilized at all levels? This notion of the multilayered "capacity" of connections is leading to a complex analysis of how compartments in society are structured, how they fit together, and how they stretch across a globalized world. These are issues that the contributions in this volume address in relation to the role of connections in African situations in the present and the past.

Contributions

The chapters in this volume consider the ideas set out in this introduction. Danielle de Lame takes the discussion on former studies of the network theory further and situates the connection debate better in this domain by emphasizing that, to understand the new developments that we are observing, network analysis alone cannot provide all the tools we need. By discussing the use of mobile phones in rural Rwanda, she demonstrates the effect that the availability of telephone connections has had on kinship networks. These connections appear to have the power to control and shape networks in a particular fashion and kinship relations, particularly distant ones are becoming much more present in day-to-day affairs. People are finding this advantageous in some cases while in others it can be detrimental. She argues that for a better understanding of the power dynamics in the "new" Rwanda, research should refocus its orientation toward these rapidly changing connections.

It is for this very reason that we invited various researchers with a long history of fieldwork to contribute to this book. They do not only present recent empirical work but they also link their findings to the past and this time perspective will help to analyze the connections. The chapters are all based on empirical work, either archival or research done in the field. The notion of the field needs redefining if we want to understand linkages, connections, and what form the linking is taking. We have to go beyond multisited research to understand the connection and the connecting. Inge Brinkman and Mirjam de Bruijn reflect on these methodological issues, questioning the methodologies they themselves developed in the "Mobile Africa Revisited" research program. Central to their research is how the "mobile margins" are being (re)shaped by new means of communication, that is, the mobile phone and new infrastructure in previously underdeveloped regions. They discuss the methods of transport communication and the mobility of the researchers themselves as not only a means of facilitating research but also a way of grasping connections.

The mobile phone is a good example of what we have called material connections, which, like a bridge, we can observe in terms of its physical presence. These material connections often interact with others such as roads, cars, and railway lines but also with immaterial connections such as those in organizations, institutions, and partnerships. All the chapters show these interlinkages and discuss how they have shaped changes in connections. The interlinkages have a temporal aspect as they have appeared in certain historical periods and then continued to exist or have disappeared. While connections have a lived history, so too do the interlinkages between them. The chapters by Walter Nkwi, Jan-Bart Gewald, and Neil Parsons view connections from a historical perspective. Parsons describes how the construction of the telegraph from South Africa to the territories to the north meant not only that a new infrastructure was put in place but that with this new connection, new political, economic, and social structures came into being. This new technology also changed ideas about progress in the future as expectations of this symbol of modernity were high. Churches in Cameroon have had a similar effect. In addition to bringing a dialectic with a world religion, a connection to a world of reading and writing and to a new infrastructure of education, medical care, and moral authority, they also brought ideas that have been important as connecting and disconnecting devices, as is shown in the contribution by Nkwi. The church gave women the chance to escape the strict regimes of chiefly authority and liberate themselves from the yoke of patriarchal obligations but to do so they needed to demonstrate a new connectivity. Signaling a connection to a church made possible a range of new relations in terms of their position, their economic activities, and their membership of nonkin social settings. Jan-Bart Gewald's chapter shows that people and their ideas can be connecting devices too. He examines the role of a British colonial officer in southern Africa who connected his knowledge of local, rural situations to Western schemes of agriculture, production, trade, and exchange and, by doing so, changed the geography and sociology of the area forever.

These historical examples demonstrate how the processes we observe in the present have their roots in the past significance of connections, but a significance that cannot just be studied and understood retrospectively. A major difference with these past situations is that today's possibilities for connecting are more numerous, but connections and new possibilities of connecting need to be recognized in terms of existing ideas of value. Some will be recognized in view of their history, while others are new and need to be freshly interpreted. New appropriations of connections give them a social life, much as Appadurai described for objects and commodities. Connections will shape and reshape society as society reshapes connections.

In this light, the contributions by Rijk van Dijk and Astrid Bochow discuss social relations as a form of appropriation of connections and their changing valorization. Both authors analytically distinguish connections and social relations in case studies of weddings in Ghana and its diaspora (Bochow) and in Botswana and its urban and peri-urban environments (van Dijk). Due to the intensification of connections, the moral framework and expectations of marriage are higher and the new possibilities for connecting are turning relationships into a financial, social, and emotional resource and sustaining or augmenting the value of the relationship. The social life of marriage as a relationship and marriage as a connection need to be explored, and while a marriage can become disconnected due to international travel, it may bring a new valorization of how connections between a couple become important and lend new meaning. A relationship can be an emotional, economic, and social resource and connections can transform in this regard too. This process goes in both directions: people might capitalize on the institution of marriage and turn it into a business enterprise, while at the same time emotionalizing it. This also demonstrates the everyday reality of the multilayeredness of connections. Married couples connect in several ways, and Bochow shows how certain forms of connections become more meaningful across (large) distances than others. Yet, while certain connections are strengthened (phone calls, sending money), other disconnections at other layers of social reality take place. Wedding ceremonies in Botswana, though using ritual forms and objects of connectivity, can disconnect people from kin, friends, and access to certain economic resources, as is shown in the contribution by van Dijk, leading to a specific compartmentalization in social life on the basis of the "activation" and "deactivation" of certain connections as compared to others.

The examples of connecting organizations and institutions, such as the public-private partnership (PPP) model (Southern Africa) in the chapter by

Marja Spierenburg, the Islamic NGOs in Chad in Mayke Kaag's contribution, and the love ritual in urban Mozambique in Linda van de Kamp's chapter, are explicit in their effect on restructuring social relations and their contents. The multilayeredness of these relations shows itself again in these cases. These organizations all work as bridging devices and encourage new relations that would not have been possible without the new option of these organizational frameworks. The connections that they forge clearly become a resource in themselves. In the PPP partnerships as well as in the Islamic NGOs, the resource is represented by its financial connections and the access that people have to others who can help them. The organizations evoke new ideas about the future and even come with a completely new rhetoric about development in society. In a country like Chad, which has experienced more than 40 years of civil war, expectations may turn into disillusionment that leads to new disconnections. The love ritual in Mozambique, however, has had a discourse of progress by using emotionality as a resource for a successful relationship for the individual. This is closely related to the message of African Pentecostal churches. The more people are prepared to emotionalize their relationships by publicly connecting their feelings to a programmatic and institutional framework that the church has established, the more their relations will benefit, or so they are promised.

The notion that connections become resources, especially in their social relating, that is, social capital, is present in the cases discussed in the chapters on Nigeria by Obono and Obono, on Angola and Cameroon by Mirjam de Bruijn and Inge Brinkman, and on Rwanda by Danielle de Lame. All of them take recent developments around the introduction of the mobile phone as an example of the bridge or connection. However, the resource is not the phone itself but the possibilities it opens up to connecting to others or to deliberately disconnecting. Not everybody can use this resource and access is a politicized issue in all cases. However, as the case of Lagos shows, the cell phone has been appropriated as social capital by a large majority of the population today and it is making people relate in new ways. Another important aspect of this connecting is also demonstrated: how these means of communication are opening up new economic worlds and possibilities, even for the poor and the marginalized. Whether this will indeed lead to a repositioning of social groups is still questionable. It seems that marginalization can also be reproduced and intensified by this new means of communication, leading to a world with large global shadows (Ferguson 2006).

These case studies plainly demonstrate the new compartmentalizations of society, where the significance of distance in the establishment and continuity of relations is losing its importance. The choice to relate has no limits. Wouter van Beek presents a clear example of this compartmentalization. The Dogon in Mali, known for their tourist encounters, are organizing themselves in the virtual world and the Internet has become their highway to communicating and selling themselves to the world. The Internet is now the forum where identities and ethnicities are discussed and recreated. Ethnicity is becoming a tool for distinguishing oneself from others but it does so mainly because it has become a resource (tourism). It is difficult to assess the consequences of the reinvention of ethnicity in this virtual space.

The process of compartmentalization may have huge consequences for the position of immigrants in a society. By keeping in touch with people at "home," they may have problems integrating in the new society. The flip side is that this will be an even stronger incentive for perceiving those who come from outside as strangers and for not sharing resources with them, thus denying them access to nationality and legality. In the literary chapter by Francis Nyamnjoh, who rethinks the positions of the Makwerekwere in Botswana, this problematic is shown in all its disturbing images of disconnections and violence. Presenting his chapter in the form of a novel based on in-depth research, Nyamnjoh (2006) brings us back to the discussion on methodology and (re)presentation. It demonstrates how academia can connect to a novelist representation of people's daily lives.

This leads us to emphasize once again that the world with all these new bridges and connections may not be the peaceful world that is the implicit promise in the theories on information societies. However, as Daniela de Lame shows in her review of these theories, the world will have to face up to more separation and schisms and less conviviality than ever before. It is time to recognize the new compartmentalizations of our world.

Notes

1. http://www.saintlouisdusenegal.com/english/index.php.
2. For a discussion of agency, see de Bruijn *et al.* (2007).
3. See Granovetter (1973) on strong and weak ties.

References

Appadurai, A. (ed.) (1986) *The Social Life of Things. Commodities in Cultural Perspective.* Cambridge: Cambridge University Press.

Appadurai, A. (1995) "The Production of Locality," in: R. Fardon (ed.), *Counterworks. Managing the Diversity of Knowledge.* London: Routledge, pp. 204–225.

Castells, M. (ed.) (2004) *Network Society, A Cross Cultural Perspective.* Cheltenham: Edward Elgar.

Castells, M., M. Fernández-Ardèvol, J. Linchuan Qiu & A. Sey (eds.) (2007) *Mobile Communication and Society, A Global Perspective.* Cambridge, MA: MIT Press.

de Bruijn, M., R. van Dijk & J-B. Gewald (eds.) (2007) *Strength beyond Structure. Social and Historical Trajectories of Agency in Africa*. Leiden: Brill.

de Bruijn, M. & R. van Dijk (2012) "Connecting and Change in African Societies. Examples of 'Ethnographies of Linking' in Anthropology," *Anthropologica* 54: 45–59.

Feldman, G. (2011) "If Ethnography is more than Participant-Observation, then Relations are more than Connections: The Case for Nonlocal Ethnography in a World of Apparatus," *Anthropological Theory* 11 (4): 375–395.

Ferguson, J. (2006) *Global Shadows. Africa and the Neoliberal World Order*. Durham & London: Duke University Press.

Glick Schiller, N. (2005) "Transnational Social Fields and Imperialism: Bringing a Theory of Power to Transnational Studies," *Anthropological Theory* (5): 439–461.

Granovetter, M. (1973) "The Strength of Weak Ties," *American Journal of Sociology* 78(6): 1360–1380.

Green, S. (2002) "Culture in a Network: Dykes, Webs and Women in London and Manchester," in: N. Rapport (ed.), *British Subjects: An Anthropology of Britain*. Oxford: Berg.

Guyer, J. (1995) "Wealth in People, Wealth in Things," *Journal of African History* (36): 83–90.

Hannerz, U. (1992) *Cultural Complexity. Studies in the Social Organization of Meaning*. New York: Columbia University Press.

Hannerz, U. (1996) *Transnational Connections: Culture, People, Places*. New York: Routledge.

Lefebvre, H. (1991) [1974] *The Production of Space*. D. Nicholson-Smith (trans.). Oxford: Basil Blackwell.

Masquelier, A. (1992) "Encounter with a Road Siren: Machines, Bodies and Commodities in the Imagination of a Mawri Healer," *Visual Anthropology Review* (8): 56–69.

Masquelier, A. (2002) "Road Mythographies: Space, Mobility and the Historical Imagination in Postcolonial Niger," *American Anthropologist* 29(4): 829–856.

Nyamnjoh, F. B. (2006) *Insiders and Outsiders: Citizenship and Xenophobia in Contemporary Southern Africa*. London: Zed Books.

Portes, A. (2000) "The Two Meanings of Social Capital," *Sociological Forum* (15): 1–12.

Putnam, R. D. (2000) *Bowling Alone. The Collapse and Revival of American Community*. New York: Simon & Schuster.

Strathern, M. (1987) "Producing Difference: Connections and Disconnections in Two New Guinea Highland Kinship Systems," in: J. F. Collier & S. J. Yanagisako (eds.), *Gender and Kinship. Essays towards a Unified Analysis*. Stanford: University of Stanford Press, pp. 271–300.

Strathern, M. (2004) *Partial Connections*. Savage, MD: Rowman & Littlefield (1991). Re-issued by AltaMira Press, Walnut Creek, CA.

van Binsbergen, W. M. J., & P. L. Geschiere (eds.) (2005) *Commodification. Things, Agency, and Identities (The Social Life of Things Revisited)*. Munster: LIT Verlag.

CHAPTER 1

Flows and Forces: Once Contained, Now Detained? On Connections Past and Present in Rwanda

Danielle de Lame

Our societies are constructed around flows: flows of capital, flows of information, flows of technology, flows of organizational interactions, flows of images, sounds and symbols. Flows are not just one element of social organization: they are the expression of the processes dominating our economic, political, and symbolic life....Thus, I propose the idea that there is a new spatial form characteristic of social practices that dominate and shape the network society: the space of flows. The space of flows is the material organization of time-sharing social practices that work through flows. By flows I understand purposeful, repetitive, programmable sequences of exchange and interaction between physically disjointed positions held by social actors.

(Castells 1996: 312)

Introduction

Nobody would pretend that Castells has ever been an Africanist. This quote could apply to all societies and could even be heard as a Levi-Straussian echo of *La pensée sauvage.* In a more casual way, pointing to the creation of language subcultures as a way of channeling communication and diverting its flows, a Kenyan student of language creolization added: "If we did not construct Africa as very special, our object of study would vanish." The regulation of flows within and between societies seems universal. Constructing representations and giving legitimacy to their regulation, however equally universal as a process, happens in ways specific to societies and embodies their values.

Perceiving flows in capital-driven societies goes far back in time. When it comes to African and other societies—such as the most publicized *Trobriand Islanders* (Malinowski 1922)—where money played a minor role but where the perception of a cosmology took pride of place, pointing to the circulation of flows is a more recent trend. In such instances, an analytical accent on metaphors to the detriment of the observation of metonymic manipulations has obscured the connections between ideology and the material circulation of goods, overshadowing the fact that economy and power are intrinsically connected with materiality. Dynamics of flows as life were probably first studied by de Mahieu (1985), who indicated the necessary, bodily enclosure of flows. Devisch (1987) insisted on the correlate enclosure, as he discerned a homologous construction of space in habitation and other humanized spaces. Taylor (1988: 1343–1348), who relied on the published text of the Rwandan kingly rituals (d'Hertefelt & Coupez 1964; de Lame 2005b) and on observation of local therapeutics, stressed the importance of flows but failed to analyze the correlate, identity-building circumscription of space by flows.

As synthesized by Strathern (1996), anthropology is not only about enclosed spaces, it is also about flows and hybrids. This chapter is an attempt to trace changes in connectivity in Rwanda and the hybridization of connections over time. Pointing to the construction of enclosed objects and authentic cultures by anthropologists, Meyer & Geschiere (1998: 603) indicated the concomitant process by which African people cope with potentially overwhelming, yet socially restricted flows of globalization, and local attempts to regain jeopardized identities. Their perspective can be applied to the old Rwandan conception of space and identity defined by the observance of ritual practices, with a highly ritualized use of space being a reflection of a world vision. There is a Rwandan saying, *Kuba mu Rwanda ni ukwizirira*: "being in Rwanda is a matter of observing ritual prescriptions," implicitly, more than a matter of borders. The will of the king restricted the circulation of people as much as spiritual forces circumscribed the community of Rwandans. Restrictions on the right to vote during the recent elections show that the idea of the "good Rwandan" based on social practices still applies.

Performed in community and kingship rituals, Rwandan cosmology circumscribed the mystical space of the kingdom to the well-ordained circulation of an impersonal force (*imana*) dispensed through the king and reinforced by the people through ritual metonymic testimonies of faithfulness to their worldview in everyday life. Shared identity embodied in similar practices circumscribed the space. Power was defined by the flow of body fluids, people, and goods, with their circulation inscribed in a landscape of dispersed habitat. This consensual ideology, reflected in the conception of body, health, and fecundity in a lineage-based society, materialized in the spatial organization of the house and in the ritual organization of the kingdom. Translated into the idiom of constant expansion by the Nyiginya Kingdom, this ideology became instrumental in legitimizing the supremacy of the royal lineages while masking the networks of forced imposition and power rivalries (Vansina 2002). The old kingdom expanded, shook, then consolidated under colonial rule before it collapsed politically, with its ideology remaining alive until its genocidal implosion. It was then superseded by a the victors' capture of connectivity, which amounted to a monopoly of access to extraverted flows. These ties have become the main warrants of survival for a part of the Rwandan population that is favored under the neoliberal paradigm of global economy.

To summarize, two modes of connectivity became intertwined over time: one based on "Rwandanness," the sharing of beliefs and the related ritual practices that constantly reaffirm social links, and the other, at once bureaucratic and extravert, based on power and esoteric communication, which eventually relinquished the last remnants of the local worldview to the kaleidoscopic dustbin of social memory. In the first part of this chapter, the old worldview and its changes are depicted and then this mode of connectivity is contrasted with the current state monopoly on communication.

Containing Flows: A Matter of Life

Two orders intertwined in the organization of connectivity in ancient Rwanda. One was inherited from the old lineage-based ideology and, as Vansina (2002) and others stressed, this provided the basis for kingly rituals. The other mode was closely linked to the expanding power of central Rwanda and the accompanying administration based on clientship (Newbury 1988). While their interwoven fabric loosened over time, the two ideologies coexisted until a final blow projected Rwanda into the impersonal order of globalized economies. This was done by imposing a state hegemony in communication, partly in tune with the old court ways, at the cost of dismantling close ties based on physical proximity. How this transformation in connectivity could have happened so late and then so quickly can be accounted for by facts we have to briefly recall.

Rwanda, a small country in eastern central Africa, was the last African polity to be entered by Europeans. Landlocked, it was also hostile to penetration (Stanley 1890: II, 332), and contacts with foreigners, including those that happened during what little commerce there was in salt (Stanley 1890: II, 315) and textiles with neighboring countries from the nineteenth century onward, were closely supervised by the court and its appointed chiefs (Linden 1977: 21). With Belgians on its western border, Germans to the south and east, and the British to the north, Rwanda was to join the bulk of colonized nations on the eve of the twentieth century. The visits of German envoys, from 1896 onward, provided Rwandans with their first opportunity to meet Europeans. Colonizers stabilized the power of Rwanda's monarchy within the borders they drew.

As noted by Reyntjens (1985: 95), the current Western notion of national territory does not match the Rwandan one, with several factors connoting various types of integration. The people inhabiting this area shared the same language and culture, while their political ties with the court varied depending on the region. There were no villages, and habitations dispersed over hilly slopes, paths leading to the chiefs, to markets, and to the administrative and church quarters reflect the intensity of communications. The enduring inscription of social ties on the landscape that welcomed people's steps would not be understandable without accounting for the ideological order making sense of geography and history. Local views on connections between this world and spiritual forces legitimized and facilitated control by the court, as its power expanded to cover a territory roughly equivalent to today's Rwanda, with degrees of integration varying according to historical circumstances (Liebhafaky Des Forges 1972: 1–32; Newbury 1988). Natural borders, with Lake Kivu to the west, volcanoes to the north, dry savannah in the east, and swamps in the south, made it easy for the royal court to incorporate geography into the ideology that justified the dynasty: life was being perpetuated thanks to a well-ordained—and contained—circulation of supernatural forces (*imana*) that the king (*imana*) conveyed to his land with which he identified mystically (Maquet 1954: 147). Rituals marked the territory and conveyed ritual efficiency to royal moves (d'Hertefelt & Coupez 1964). Geographical elements, such as the Nyabarongo River, divided the territory in the same way that a belt would divide the body, and determined the moves of the king on his territory: peaceful kings, defined as such by their order in the dynastic succession, were not supposed to trespass its course. Trespassing would endanger the king and the kingdom.

Locations of ritual significance hosted the kingly peregrinations, allowing for political control in a fluid space (Liebhafaky Des Forges 2011 [1972]: 11). When the levy of "provisions" for kingly sojourns in various residences merged into a permanent levy, the first polarization of the connective space was achieved. Bringing the annual levy (*ikoro*) was an occasion for the people to meet. Except for the king's or the chief's courts, where clients and commoners could meet and exchange news, the (few) markets were the only places of converging errands. Otherwise, links were numerous but mainly local (Meyer 1911), coinciding with material exchanges of goods for practical or ritual reasons. Simple exchanges of food and drink were tokens of good neighborhood relationships, while solemn exchanges within the context of matrimony, clientship, and blood brotherhood built strong links that increased the power of those involved. Power translated into the number of people you could rally to your cause, into your ability to connect. Alliances were of paramount importance for the sake of security in a context of dispersed habitat and perpetual intrigue, and they materialized in visits during which news and gifts were exchanged. All news was, of course, interpreted according to local knowledge and circumstances, and rumor thrived. Usually, as Linden (1977: 43) noted, "a peasant had little access to information beyond his own group of hills" unless he had contracted alliances that would bring him to travel, go in search of new land (with new settlements inducing family visits and new alliances), or go into exile. Connectivity was restricted by the chief's control; spies were instrumental. The split in command over a similar area (with chiefs of the army and chiefs of the cattle competing for power over overlapping domains) ignited rivalries and fostered intrigues. To prevent falling victim to these, chiefs would spend long periods at court and their followers would accompany them for a time, bringing news as they replaced each other. Markets developed near the borders and at meeting points of different agricultural zones, and they became places of high connectivity. Not all of them were under the direct control of the court but of lineage heads; however, local specialties, such as the annual levy paid in kind, would reach the court.

When Europeans arrived, Rwandans initially retained the mastery of contacts, including visual contacts that were supposed to convey potentially dangerous forces (Ntezimana 1980; Kagabo 1995; de Lame 2005b) until they reluctantly allowed Europeans to settle. Missionaries were to follow the Germans, who would be replaced by the Belgians in 1916. Under colonial rule and with the capital in Nyanza, the royal drums stopped sending messages to people surrounding the many ritual sites (Smith 1997). Eager to strike a blow at "superstitions," and aware of the mighty rituals, missionaries forced the king to transgress the rule and cross the Nyabarongo River. However, this did not suffice. They took care of having the king deposed by the Belgian administration and ensured that he was replaced by a more complacent successor in 1931. The imposition of foreign rule, the transformation of the administration, and the close collaboration between administrators and missionaries created multiple polarized spaces and new modes of connectivity that intertwined with previous ones. A deliberate policy to keep Rwanda rural was successfully applied (Sirven 1984) until 1994, with 95% of the population still living in the rural areas at this time. This allowed for the continuation of lineage-based modes of connectivity along with the multipolar bureaucratic ones. The containment of space within cultural borders no longer matched extravert politics. The first crack, which echoed a century later with the return of exiles and the exile of a new "caseload," had been the opening up of the country.

A century earlier, indeed, observing the dense network of paths that testified to the intense but mainly local circulation of people, the Germans had seen the breaking down of the kingdom into small chieftaincies as the only way to open Rwanda up for colonial exploitation. The country would be integrated into the German Protectorate through the creation of a railway (Meyer 1914) and this would result in the export of agricultural produce. The railway was never built. During the First World War, the extraction of a so-called surplus manpower, combined with the climate, resulted in a famine (Botte 1985). Help provided by missionaries resulted in a wave of conversions, allowed for a tightening of the Belgian administration (Newbury 1991: 285), and facilitated the introduction of new crops (Botte 1985; Cornet 1995; de Lame 2005b: 52). Abundant manpower was a resource to extract and export. The Belgians exported Rwandans to Congo, where they retained a sense of identity and found Rwanda speakers in Kivu who had been living there for centuries. Exporting people enlarged the "mental territory," while taxation (in cash) and the quest for money through migration to Uganda brought a new experience of space, disconnecting ritual and political borders. The very presence of foreigners was an external reference overshadowing the local worldview. Changes in administration transformed connections: taxation on work or goods directly brought to the chiefs was replaced by a bureaucratized connection involving physical moves toward foreign lands or the quest for employment with the few foreigners, mainly missionaries and civil servants and a few merchants. The ties between peasants and chiefs became looser.

The pursuit of cash involved new allegiances: *imana* changed meaning and money changed hands. Missionaries influenced connectivity in several ways. Schools, health centers, and chapels attracted "customers" and launched new rhythms in time and space, even if slow means of transportation—walking or the odd mule—were still the order of the day. Churches and chapels were built on hilltops like the chiefs' compounds, proclaiming the conquest of souls, and drums launched new messages through codified rhythms (Smith 1997) that once belonged to the chiefs. The polarization of space/time by the state and churches was often competitive but institutions usually reinforced each other, as administration centers joined churches, schools, and health centers. As providers of employment, missionaries started monetizing the economy. Attending mass required garments that had to be purchased, and school attendance was initially paid for in salt and involved apparent complacency with the new ideology. A polarized connectivity to God replaced the man-reinforced power of the local *imana*. From the 1920s onward, missionaries adopted the methodology of "stepping stones," engrossing local notions with new meanings. A new interpretation of the word *imana* captured the diffuse power it connoted, precipitating its flux into the notion of a personal God. The bonding of people through the common observance of rituals continued to thrive as *imana* made place for a double entendre. The desecration of the royal drums through their exposure to popular gaze struck the hardest. Christian ways precipitated the condensation of fluid forces into an imported cosmography and paved the way for the extraversion of the country. Missionaries had a quasi monopoly on one indispensable medium of extraversion: writing. Writing became a condition for command. For decades, the sons of chiefs accessed quality teaching, with limited number of peasants being able to write and only seminaries welcoming Hutu youngsters. Political power was linked to new modes of connectivity but the old ones remained active in daily manifestations of sociability and sustained "weak ties." On the poorly maintained roads, cars started to be a mark of one's status. After the Second World War, the Catholic Church took a socio-democratic turn: new elites were formed and subversion of the old order got support from the Church press, culminating in the publication of the *Manifeste des Bahutu*. Traveling to Europe became a token of status, with the king traveling to Belgium and the first president-to-be training as a journalist on a Belgian newspaper. Having connections now meant international connections. Forces were to be drawn from outside, and the intrusion became manifest with Belgium's support for the revolution in 1959. Communist China supported the Rwandan monarchy.

Around this time, the national Radio Rwanda was set up under the direction of a Belgian journalist. Its inauguration took place in May 1961 with an evening program lasting two and a half hours (Rukebesha 1985: 86). With an estimated 3,000 receivers at the outset, Radio Rwanda would have reached an audience little interested in its agricultural advice (*Ibid.*). From the start, Radio Rwanda was part of the diffusion of the development ideology that became the main tool of the Second Republic in 1972. From then onward, the country opened up to increasing foreign aid. Wealth was perceived in terms of access to modern consumption, a blessing to come from external sources. Apart from visible, official channels, such as the radio, that claimed to be part and parcel of everyday governance (with an estimated 26 receivers for every 1,000 inhabitants in 1979), the local redistribution of wealth and power still followed the old ideology of connectivity.

The celebration of alliances, which had been quite private in the past, became ostentatious with exchanges of cows and the ritualized sharing of drinks that conveyed to the assembly a message of complicity embodied in the sharing of fluids, including saliva. Early missionaries had triumphed when they had succeeded in having King Musinga's uncle drink with Christians (Kayihura 1950). The implications of this gesture become clear when one realizes that, under ritual circumstances such as divination or sacrifice, the Rwandan word for saliva would not be *amacandwe* but *imbuto* (seed). The very substance of the guests was being incorporated by all, as cleaning the straw of one's neighbor's saliva was considered an act of sorcery, a break in the circulation and an appropriation of a vital substance for one's own purposes. With increased ostentation and a connotation of neo-traditionalism in the Second Republic, this vital mode of connectivity became a binding tool akin to the long-abandoned blood brotherhood sealed by the drinking of each other's blood. With the Second Republic's commensality, a broader peasant congregation was created following a similar semantic pattern: the display of sharing sealed and recorded a commitment to an agreement, the content of which was known only by a limited congregation. The inclusion of modern drinks in festive assemblies and the abstention from alcohol by Born-Again Christians compartmentalized the assemblies along the lines of distribution and conveyed multiple messages (de Lame 2005b: 306–340). Festivities organized by high-ranking politicians made connections obvious and displayed the close links between the top and their lineage-based country-side connections. Bureaucracy followed the trails of the lineage connectivity that had survived the imposition of the central kingdom's rule and its colo-nialist avatar. With ethnicity resurfacing as a crucial discriminatory element in the late 1980s, assemblies became increasingly monoethnic, with the Tutsi minority spread over the territory being excluded. The thinner networks of the Tutsi people made for habits of longer-distance communication in line with the links of chieftaincy in seemingly bygone times.

This kind of links had persisted under the two first republics along the paths of commerce and administration. Modesty had been a publicized virtue of the First Republic, with the president driving a small VW car. The incon-venience caused by dirt roads was compensated for by small airplanes (*petits porteurs*) that would transfer their well-to-do and well-connected passengers from one part of the country to another, especially Kigali, the capital city, and the university town of Butare. The bigotry of the First Republic is in sharp contrast with the opening up of the country to foreign aid under the Second Republic and the presidential authorization to civil servants "to participate unrestrictedly in productive enterprises" (Bézy 1990: 41). Lineage-based net-works needed improved road transportation. Roads had first served colonial interests in the past, as they connected Kigali with Bujumbura in Burundi, the colonial capital of Ruanda-Burundi at the time, and with Kenya, especially Mombasa, from where fuel and other goods came. In 1976, the very few vehicles traveling in Kigali had to reckon with the country's first two traffic lights. Industries developed little: trade coupled with official employment and country connections (for food and cheap manpower) were the scanty foundation of a capital that would allow for more trade, the acquisition of trucks, and investment in housing for rent. Wealth in cash proved highly connective, with a small class of wealthy civil servants nurturing a class of merchants with international connections and East African ramifications, while back in their rural place of origin, they still displayed their wealth in the ceremonial exchange of cows that enhanced their status.

Until 1988, any appointment had to be made by letter or radio or, as in the past, by sending a messenger. The telephone was to change this. Its popularity spread fast among people who had access to electricity, a tiny minority located in administrative centers and in small concentrations of houses that could barely qualify as towns. This lack of connectivity partly accounts for the persisting reference to experiences of time/space that were strongly embodied and linked to concerted social rhythms that included, for a fraction of the rural population, listening to the radio, its personal messages, and the president's discourses. Owning a radio and the batteries to use it was still the privilege of a minority in 1989 (5.5%), a token to carry in processions at peasants' wedding ceremonies, broadcasting religious cassettes from under protective crocheted napkins in bright colors. To some in remote areas, radio was also a means of connecting with the wider world, via Radio France International (RFI) and the British Broadcasting Company (BBC) World Service. The choice of sender was the "horizon" (Hannerz 1992) of the listener. This led my research assistant, a local farmer who had only had a few years of secondary school, to comment on the fall of the Berlin Wall: "Africa is finished: donors will not care for us anymore but concentrate on Eastern Europe instead." With his desire to enlarge his horizon, his choice had been not to be a prisoner of the small world of the hill where he was born. As a member of a nonconventional church that had opened up different perspectives for him, his choice made him a local cosmopolitan.

The international response to the opening up of the country by President J. Habyarimana had been enthusiastic. In about a decade, the Second Republic produced a middle class composed of civil servants and their family networks (Bézy 1990). The geopolitical reshuffle came at a stage when donors were disappointed by the results of their development policies and considered peaceful, peasant, Christian Rwanda a field where to restore their image while enjoying a mild climate. Money and expats flocked to Rwanda and the telephone followed. Soon disillusionment started to taint considerations about the results of cooperation with Rwanda (Hanssen 1989) while the World Bank imposed its neoliberal catechism and structural adjustment programs (SAPs). The new middle class closely linked to central power ramified along the lanes left open by the lineage-based ideology. Connectivity, materialized in local cosmopolitanism, reflected power and the economic divide. A vehicle could be a token of accumulation: the owner was respected for his wealth and competence and embodied a link to town. Official vehicles at private venues took on other overtones, highlighting political connections that allowed for transgression, and reinforced the link between connections and material connectivity.

Cosmopolitanism relies on the capacity to assimilate the other's ways, something expatriates were not doing as they imposed their ways on the people they pretended to develop but without any basic knowledge of their lives. Bart (1984) and Guichaoua (1989) were the first to cross the cultural border and study rural Rwanda out of the exoticism that had prevailed until then. Development projects were still being implemented blindly. Decentralization translated into attempts at polarizing rural space. Administration centers aimed at attracting concentrations of houses but only attracted small shops while people continued to live in their dispersed habitats. This strengthened the power of the local elite, draining the money of development projects along clientelist tracks, according to patterns that SAPs and multipartyism would exacerbate and canalize on in lethal ways (Uvin 1998; de Lame 2005b). Rwandans and expatriates communicated little, except on professional grounds where retention of knowledge was part of the game. The two groups lived side by side, divided by race and the memory of colonial times, and by differences in wealth and salaries. The car driven by development workers and other expatriates would have a plate marked with the letters IT (International Transit) that was translated jokingly into the Rwandan language as *Icara turiye* ("sit down so that we can eat"). Language was another barrier. Apart from the missionaries, few expatriates would learn the local language, Kinyarwanda. Rwandans would play on their language's reputation for being extremely complex to discourage foreigners from learning it. Using the grammatical complexities of the language to make oneself understood by only a fraction of one's audience is very much part of the local culture; it used to be a skill young men would learn at a chief's court as part of their training. Women would do the same more informally in this male-dominated society. Applied more casually to daily communication, intelligence (*ubwenge*) (Rukebesha 1978; Crépeau & Bizimana 1979[1]) would close off the understanding of speech to outsiders.[2] The study on the "media of genocide" can best be understood within this habitus of restricting flows of information and sharing knowledge between circles created by the use of language and the restricted access to material means of communication.

The liberalization of the press and freedom of expression—the deregulation of flows—opened up avenues for the creation of new networks and for the public expression of racist stereotypes in pamphlets that would reach the country's remotest areas. Along the trails of the civil service, of merchants' transportation, and of the transit of people with an occasional job in town, this propaganda reached the countryside. Until then, freedom of speech had been restricted to private exchanges with an overflow of usual restrictions a token of impending disorder (de Lame 1997). Its bearers responded whole-heartedly to the official ideology of "transparency" (de Lame 2004) that was the mood and the motto of the time. They were endowed with the prestige of modernity and appealed to the youth. Half-coded messages, usually not translated exactly into French (then the second national language), conveyed news that would be understood in different ways by radio listeners. This com-munication context was fertile ground for the unruly insinuation of rumors building on individual fear and fostering a protective, collective response to mobilization. After the invasion by the Rwanda Patriotic Front (RPF) from the north in October 1990, the country was divided, with the RPF army occupying a large part of the territory, causing massive displacement and dividing the internal population further by its propaganda campaign aimed at young Tutsi. Camps of displaced persons were ideal ground for the recruit-ing of militia. The young unemployed on the hills were receptive to messages that were germinating in a bed of frustration and fear. Connectivity followed ethnic channels and political divides along the lines of personal proximity; face-to-face communication took over when mass media were not accessible.

What voices dominated this grassroots innuendo? Audiences to radio broadcasts had grown over the years, soon reaching the limits of a market restricted by money shortages. The ownership of radio receivers grew to a ratio of 55 to every 1,000 Rwandans in 1989 (UNESCO statistics quoted in Higiro 1992), with a concentration of owners in towns and among money-earning rural people. Purchasing batteries often remained problematic. The radio was a government instrument that followed the rise of a personality cult and conveyed the development ideology, with journalists relaying messages and reporting on local events (Rukebesha 1985; Higiro 1992). No detailed study of the effects of the radio on Rwandan audiences exists (Rukebesha 1978, 1985; Bart 1984; Lenoble-Bart & Tudesq 2008) but radio broadcast-ing was the first way of bypassing spatial constraints and the circulation of persons. It conveyed messages that were centrally screened, with inconsisten-cies that left space for contradictory interpretations. This was the background for the opening of nongovernmental radio stations, such as the brutal Radio-Télévision des Mille Collines (RTLM) that would deliver messages endowed with the aura of modernity and freedom. Chrétien (1995) gives an account of the broadcasts in the context of the 1994 genocide though no a empirical assessment of its effects in rural areas exists.

Radio Rwanda was the voice of authority, a governmental agency in a de facto single-party state that rested on the personal power of the president and his wife, with its educational (health, nutrition, and agriculture) and recreational programs (popular theater and folk tales) spreading a modernity in tune with the government's development ideology. This made it the voice of a seemingly benevolent authority dedicated to the well-being of the population. Rural farmers often referred to the sending of messages through this media as resorting to *ibitega,* magical ways of influencing someone's will from a distance. Broadcasting excerpts of presidential addresses every day at lunchtime started to talk about a state jeopardized by the side effects of liberalization and by the offspring of exiles who had settled in Uganda after the 1963–1964 genocide. Pamphlets paved the way to genocide with hate messages. Off the beaten track, newspapers spread the president's dismissal of the Arusha Agreement as a "scrap of paper," while L. Mugesera launched the first call for ethnic cleansing. Harmony, however, remained the order of the day in radio talk shows while news of the war, attacks, and arrests and hardened political polarization contradicted the motto. From behind the curtain of human rights, the then-director of Radio Rwanda, F. Nahimana,[3] launched the first incitements to mass murder (Chrétien 1995: 56–59). Democracy became applicable only to the majority, to the exclusion of the (15%) Tutsi minority. Under the paradigm of local cosmology, unruly connectivity fitted well in the representation of an ominous disorder that called for a restoration of order and the enthronement of a new king (de Lame 1997). Another omen was the invasion of the country by the RPF under the leadership of Paul Kagame in October 1990. The international community tried to foster negotiations and the embarrassing Nahimana was dismissed from Radio Rwanda, which he then accused of "serving the enemy."

The split between the community of often naive expatriates and the Rwandan elite was going to serve his project. The idea of a free, populist radio had been around for a long time. The liberalization of the media was applauded by the West and the demands for a populist channel could be met. Radio-Télévision des Mille Collines became the best-known medium for the propagation of genocide orders. Incitements to action or to work (*gukora* was the word used and is quite flexible in its connotations) were broadcast alongside popular modern music. The local elite played a decisive role in relaying RTLM. Literature abounds on the interplay of mass broadcasting and mobilization on the basis of personalized links based on lineage and neighborhood connectivity. Proximity was not sufficient to ensure protection (Fujii 2009; Longman 2010). The churches, thanks to their regular audiences, could have played a role in discouraging the genocide but they did little (Saur 2004; Longman 2010) and priests often joined in the deadly propaganda and genocide. Looking at the practical aspects of the genocide's implementation from a cultural standpoint is a dubious exercise in which I will not take part. Roadblocks seem the most efficient way of checking identity and filtering people along ethnic lines. Yes, flow was being severed or blocked. Many ways of killing were used, however, and not only by the Hutu militia. I will put such issues to one side here and will concentrate on two complementary aspects of Rwandan policy: the destruction of peasants' connectivity and the hegemonic production of knowledge.

Detaining Flows: A Matter of Hegemony

The literature on the genocide of Tutsi in Rwanda and the other massacres that preceded and accompanied it is abundant and diverse in quality. It becomes contradictory when it talks about the achievements of the current government and this has a lot to do with the management of information flows. Ingelaere (2010) has underlined the quasi-opposite images top-to-bottom and bottom-up reports and research produce. As Ingelaere stresses, the authors, often in good faith, advertise the image the government promotes as they restrict their contacts to Kigali and the elite, while reports that could challenge what he aptly calls "the aesthetics of progress" are kept out of sight (*Ibid.*: 46–50). Barely known before 1994 (de Lame 1996: 3–12, 42–44, xx), Rwanda is now the nexus of contradictions resulting from propaganda and a top-down approach (Pottier 2002; Ingelaere 2010: 41–45).

During the genocide itself, material conditions of communication made it almost impossible for journalists to account for the horrors that even the hardened specialists among them could barely cope with, let alone describe (Dowden 2004). Even reliable studies have framed aspects of the genocide to the detriment of an encompassing picture of the political stakes in the region. Even the Human Rights Watch account "Leave None to Tell the Story" (Des Forges 1999) has recently been pointed at as a framing that may have served the dominant vision propagated by the United States (Herman & Peterson 2010). The autobiography of Umut4si, who is not the only one to account for RPF atrocities (Prunier 1995; Umutesi 2004 Ruzibiza 2005), testifies to the crimes against humanity committed by the Rwandan Patriotic Army (RPA), while the word "genocide" only applies to the genocide of the Rwandan Tutsi. The invasion of Rwanda, the genocide and other mass killings, and the reorganization of the country have highlighted the mastering of connectivity as a key instrument of power. The New Rwanda (*Rwanda Rushya*) puts information and communication technologies (ICT) to the services of a centralized conception of power akin to that of the old kingdom with worldwide connectivity, rather than supernatural forces, a grant for legitimacy in tune with Western interests. The transformation has been fast, radical, and done in such a way as to ensure it is final. The process takes its full meaning within local conceptions of power. While popular conceptions of connectivity binding people along blood ties are efficiently fought against, the ideology of conquest that enlarged Rwanda on the eve of colonization has prevailed in the imposition of central rule under the internationally accepted guise of development and privatization. International support straddling feelings of guilt and the US need for a gateway to Congo's natural resources and an ICT hub in Eastern Africa[4] are the ultimate sources of power. But as the Rwandan saying goes, "Only the maker of the drum knows its contents," only the king knows the secret of his power.

Territory and Connectivity

According to local views, territoriality was defined by the extension of ritual practices of two orders: practices binding lineages to their ancestors and land and practices legitimizing central power. Paying tribute to the king, a token of allegiance that could fluctuate with political domination, had overtones of loyalty fitting with cultural practices spread beyond current political borders. In a similar way in 1994, the massive flow of (mainly) Hutu exiles to the Bukavu and Goma areas in Congo, Ngara in Tanzania, and Burundi expanded the conceptual territory well beyond its international borders, with about 3 million then living abroad. Serious estimates put the number of victims of the genocide and violent crime at about 800,000. About 700,000 exiles returned to Rwanda from Uganda, many bringing their cattle with them. With camps on the shores of Lake Kivu, the border was especially porous, allowing for traffic, the neutralization of witnesses, and a feeling of proximity akin to revenge. In some cases, camp settlements reflected normal patterns of social organization and connections as inscribed on the landscape back home. Remarks I heard from various sources during my own fieldwork in the Bukavu refugee camps in 1995 confirm a conception of connectivity based on shared values and habits. One element is especially worthy of mention, namely, that, while the United Nations High Commissioner for Refugees (UNHCR) felt overwhelmed (Steering Committee 1996) and called on the Congolese authorities to control the camps, Rwandans had a strong feeling of being "at home" there, seeing outsiders as intruders and any attempt at imposing a rule other than their own as an intrusion (de Lame 2005a). Mobutu's presidential guards used walkie-talkies provided by the UNHCR. Aside from this, communication between the 26 refugee camps spread over the Bukavu area happened mainly through traveling intermediaries. Refugees resorted to using the drivers of UNHCR field assistants to liaise between camps. Most field assistants were Rwandan refugees as UNHCR expatriates were reluctant to have contact with the refugee population, which they placed in the lump category of mass killers. Wealthy refugees had their own cars or the official cars they had used to flee in. Connectivity followed the familiar lines of political allegiance that were so efficient during the genocide, making it possible for the genocide's organizers to control several camps to a large extent. Estimates of Hutu returnees to Rwanda in 1996–1997 vary, with 2 million being the official figure (Bruce 2007). Social cohesion materialized within political movements in the camps and triggered action from Rwanda and the destruction of the camps. This led to refugees fleeing through Congo and to the death of many (Umutesi 2004; Tripp *et al.* 2005). This Rwandan action still needs to be assessed in the more general context of the overthrow of Congolese president Mobutu and of an informal Rwandan presence in this part of Congo rich in natural resources. Alleged Rwandan views on enlarging its overpopulated territory and the West's support for an ally that could serve its interests in this rich area may have played a role too. The consequences are still affecting life in Kivu today. Flows of people have created a Rwandan community based on common values and practices applied to diverging ends. Recomposed diasporas (Vambe 2008) over the borders have blurred under one paradigm held to similar social practices diversely performed on registers of modernity reinterpreting old themes. The main one is connectivity. Different concepts of identity reflect in the daily relations with the state, while the Rwandan state, in turn, resorts to criteria pertaining to cohesion ("génocidaire ideology," vaguely defined and versatile and which sends you to jail for several years) to label its citizens. As a reminder of the kingdom's past, Rwandanness defined as political conformity determined the capacity to take part in elections in August 2010. Prisoners and Rwandans with a refugee status in the diaspora were denied the right to vote.

Local Connectivity and Hegemony

Ibihanga bibili ntibitekwa mu inkono imwe: "one does not cook two skulls in the same pot." This proverb could apply metaphorically to the two types of connectivity I have described. The security of absolute central power excludes any other channel open to the transmission of a competing ideology of grassroots cohesion. The lineage ideology rested on kinship, alliances, and mystical and practical links to land: patriliny and virilocality combined to build a social space where extended families' compounds dotted the landscape with loosely joined enclosures. Alliances of various kinds complemented the links of kinship. The invasion by the RPF in October 1990 and the massive displacement of population resulting from its occupation of the northern part of the country struck a drastic blow for this kind of settlement pattern already weakened by land shortage. The economic significance of the family link to its land[5] loosened, though some aspects remained meaningful. A rooting in tradition, materialized in the planting of a banana grove when building a new household, had economic aspects. Banana beer channeled cash down to the peasants and was essential to commensality and strong social binding (de Lame 2004, 2005b: 306–340). In the context of land reform and a forced switch to cash cropping, banana groves are not a family business anymore. Brewing is now centralized in the hands of women's cooperatives. Banana beer had masculine connotations and drinking it with a straw, a highly charged symbolic gesture, is forbidden. Forbidden too are many practices that were part of everyday life and bound people materially and symbolically. These interdictions affect social links but also the livelihood of peasant families, while wealth concentrates in fewer hands: those who can afford to keep their cattle in sheds and produce the shoes that are now compulsory. Schemes affect cultivation in valleys, the production of charcoal, and the selling of beer. Producing an image of modernity and cleanliness goes hand in hand with monopolies on commerce, fines, and tight controls that intrude into family homes (Ingelaere 2010).

Land reform,[6] an outcome of the villagization policy introduced in 1995, became law in July 2005 and will continue to uproot rural people and change their way of life. It essentially atomizes rural society. As Des Forges (2006) noted, this reform is a further attempt by the state to imprint central rule on the use of land, which has been a long-term state endeavor (Reisdorff 1952; Vanwalle 1982;[7] Newbury 1988). Pottier (2002, 2006a, 2006b) pointed to the risks of social unrest but hoped flexibility would prevail. When considered in a broader context including laws on trade, the picture of a vast plan for a radical restructuring of rural society (Ansoms 2009) is worrying. The measures deeply affect rural life and social cohesion, jeopardize livelihoods, and presuppose a restructured connectivity resting on central control. Political tools, such as the attribution of control capacity to selected peasants and the creation of youth groups (*itorero*) taking care of conformity to government-advertised traditions, undermine social cohesion. Their monitoring is facilitated by ICT, especially the use of the mobile phone and reporting. In a context of destructured social ties, even the *gacaca* tribunals, an instrument of transitional justice and reconstruction, becomes "a politicized process: respect for survivors is overridden by questions of power" (Ingelaere 2010: 45) and marred by local greed and revenge. All these measures weaken the rural social fabric, making central control all the more effective. They apply on a remapped Rwanda breaking with the past.

In pre-genocide Rwanda, family connections binding Kigali to the countryside would have been a channel for social control by the state, while the new group in power lacks rural links. This isolation has been transformed into an opportunity: any kind of long-term connection between administrative staff and the population is being prevented, with a rapid turnover of civil servants. Reorganizing the administration and imposing easily controllable and detailed target policies (*imihigo*) put local civil servants under the tight control of district authorities and, ultimately, the president to whom they are accountable (Ingelaere 2010). Their distance from the peasantry leaves little room for afterthought and makes them dependent on the hierarchy they report to, first and instantly, by phone. Now called executive secretaries, local civil servants have a university degree and are better paid than junior university staff. This huge advantage takes care of their zeal and prevents local contact, other than on professional matters, with their cocitizens. It enables them to use the brand new roads and go back to Kigali or their place of origin on the weekend, a move exactly reverse to what happened in the past. Assisted by various committees potentially spying on them, executive secretaries have, locally, a quasi monopoly on power as long as they act as its efficient driving force. Rural teachers, a potentially binding force in the hills, can now barely afford a standard of living that ensures respectability.

Remapping Rwanda and renaming locations, to keep a distance from painful memories, imposes a rupture with all aspects of the past on all local people. This token of newness marks a gap between generations and imposes a shift on all those who have known Rwanda for a long time. Connecting the present to the past has been made as difficult as possible (Ingelaere 2010). Power is mainly in the hands of exiles returned from Uganda and the official shift from French to English in education from 2010 onward has cut civil servants off even more from rural people. Ansoms (2009) gives a detailed account of the risks involved in this restructured connectivity distancing the elite from the vast majority of the population and checking conformity with a top-down circulation of information. Communication technologies are key to this "decentralized," yet centrally controlled administration. Any minor incident in a remote area has to be reported and is immediately monitored at a higher level. Local civil servants are constantly accessible and their physical position can be controlled (Green & Haddon 2009: 28). The political apparatus rates the highest in the connectivity capital while the vast majority of the population can be qualified as "techno-marginal" in more than one way (Guazzini 2001). Internet communication is reputedly under tight government control with a "panopticon" effect reinforced by proof of its occasional reality (Foucault 1977); content filtering seems, in fact, not frequent. Accessing (critical) news from abroad is possible, but any critic of the government is forbidden, with Rwanda holding the one hundred and fifty-sixth position in the ranking of press freedom.[8] In tune with the past, keeping up the appearances of consensus is paramount.

Computers are beyond the reach of more than 90% of the population, yet ICT produces a considerable income for owners of companies who are, more often than not, officials (Nsengiyumva & Santiago 2007). It is also high on the political agenda, as shown in the government's "Vision 2020" declaration, with optic fibers linking administrative buildings in Kigali and, by 2020, hopefully a PC for every child in school (*Daily Nation*, March 20, 2010). Universities and international cooperation agencies have played a major role in fostering Internet connectivity in Rwanda. When it is used with rural teaching, however, one has to question its effect. While such an achievement would certainly help pupils to reach information (teaching at high school level is meant to be primarily technical), it will also make political indoctrination easier. The ultimate goal of ICT development is to increase the "efficiency of public services . . . through the application of e-government principles" (Republic of Rwanda 2010: 16).[9]

Mobile phones (3 million with a population of 10.5 million in 2010) have spread the net of institutional control. Can they enhance ordinary people's networks and make them competitive with the official ones? How does popular connectivity appropriate the cell phone? My main question is about the countryside with its dispersed habitats and very low incomes. I got a hint of the situation when revisiting the hill in my fieldwork area in western Rwanda. For a few years, the parish priest had had a quasi monopoly on communication, with a bandwidth so weak that he had to climb a tree to get reception. Soon, more people acquired cell phones and ostensibly used them as status symbols, while the number of masts increased for state control in the context of the implementation of the *gacaca* courts and administrative measures. At the same time, direct communication for ordinary people became easier. For example, a locally born Rwandan living in Europe could monitor the help he was providing to his native area and his family. He could efficiently resort to cell-phone technology because he knew the two contexts well, with his cell phone mainly a nonspatial embodiment of his own broad horizons encompassing European NGOs and individual benefactors, as well as the individuals locally responsible for the use of what he could provide. What I observed in 2008, apart from tight official control, was that cell phones merely facilitated existing networks. Youngsters who could purchase one were quickly restricted in its use by lack of money, except for those who had some police task, assuming that recharging the batteries had been possible at the health centre. The cell phone was an important tool for the few merchants living in the hills, allowing for a monitoring of stocks in town and for a better coordination between the two truck owners.

Conclusion

Castell's notion of flow as connection between physically disjointed actors and his "space of flows as a material organization of time sharing social practices that work through flows" can apply to metonymic practices linked to a system of representations that endows these practices with forces circulating to produce a common identity and social cohesion. While the system of reference is imaginary, it produces real effects (Hirsch *et al.* 2005: 267). The Rwandan example is a case in point. The shift in connectivity, from an imaginary reference to corporate identity and to a political ideology enacted in centripetal communication, has been greatly facilitated by the use of ICT and, in particular, by the mobile phone. This device, however, needed to combine with material infrastructure—such as roads and masts—and a reshuffle of the administration to fulfill the controlling role a centralized state acting under the guise of decentralization enticed to it. Performance targets and an image of modernity have become the instruments of a social engineering that draws its significance from the extraversion of a regime based on international support. In tune with a world craving for audit, this can be monitored thanks to modern technologies. The so-called e-government is at work there. As far as Rwanda is concerned, postgenocide policy has all to do with connections and connectivity, both in the management of knowledge and in the atomization of peasant society under the guise of national unity.

As the Jamaican study by Horst & Miller (2006) has shown, the use of cell phones enables aspects of relationships within existing networks. While the new Rwanda depends on drastic social engineering, "balancing development and national security" (Nsengiyumva & Santiago 2007), the space of flows as intended by Castells is quite circumscribed by urban space, mainly that of Kigali, the capital, and the much controlled channels of its administration. Social contacts are even more restricted to this space than business calls (Castells *et al.* 2007: 83, 283). As the first works on cities by Castells would have it, this divide is, in itself, a spatial projection of power. The rural-urban divide matches social inequalities and the incumbent restriction in the acquisition and use of mobile technologies. The channeling of flows benefits the stability of a government that can hardly be qualified as democratic, and goes hand in hand with the atomization of rural communities. This is made easier using ICT and connecting directly and instantly with those in charge of order. The Internet is easy to control and does not reach the whole country, while the use of cell phones facilitates business and administrative connections. The majority of the poor remain limited to local networks. This unequal use of mobile communication is in tune with access to the world as conceptualized by Hannerz (1992), and presents opportunities and options that are oriented toward local enterprise or, alternatively, are embedded in cosmopolitan visions of the world. It also reflects relations of power in more than one sense. Firstly, ICT are under control: messages are sent on cell phones to call for political meetings and to remind users of their accessibility. As connections are tokens of power, using these technologies reflects the extension of connections but they do so within preestablished private or public relationships. In Rwanda, the development of ICT has been linked to political power from the outset and has developed with external support, allowing for the progress of a few and for the consolidation of an extraverted hegemony. The dominant version of old Rwandan cosmology referred to a king as a conduit for fluid forces constantly vivified by the ritual conformity of all to the worldview that the builders of the and ever-expanding Niyginya Kingdom had appropriated and formalized. This force, which blessed the faithful, emerged from the source that was faith in the ways of the ancestors, *umuco karande*. Electronic flows, so easily detained for the benefit of few connected individuals, cannot match this imagery.

Notes

1. See the index in the book under the word *enge*, which is the radical of *ubwenge*.
2. The current government has mastered the cultural practice of *ubwenge* very well.
3. Nahimana was found guilty of genocide by the International Tribunal in Arusha.
4. The US Embassy in Nairobi seems to have played this role and monitored the Middle East until its bombing on August 7, 1998. The US Embassy in Kigali is now heavily protected.
5. When livelihood and reproduction are linked to the use of land and when its productivity is seen as connected to a worldview and to the rituals that sustain it, it is difficult to dissociate economy and symbolic action.
6. For a review of literature, see Mushara & Huggins (2004).
7. For French abstracts and other references, see d'Hertefelt & de Lame (1987).
8. Reporters Without Borders regards the situation in Rwanda (156th) and Equatorial Guinea (161st) as very grave because of the control that their governments exercise over the media and freedom of expression in general Reporters Without Borders 2012: 5. See also Human Rights Watch World Report 2012.
9. President Paul Kagame has headed the Intelligence Services in Uganda on the basis of his training at the US Military Academy at West Point, where his son is currently enrolled.

References

Ansoms, A. (2009) "Re-engineering Rural Society: The Visions and Ambitions of the Rwandan Elite," *African Affairs* 431: 289–309.

Bart, A. (1984) "La presse écrite et la radio," in: Fr. Jouannet (ed.), *Le Français au Rwanda: Enquête lexicale.* Paris: SELA, pp. 223–250.

Bart, F. (1993) *Montagnes d'Afrique, Terres Paysannes. Le Cas du Rwanda.* Bordeaux: Presses Universitaires de Bordeaux.

Bézy, F. (1990) *Rwanda 1962–1989, Bilan socio-économique d'un régime.* Louvain: Université Catholique de Louvain.

Botte, R. (1985) "Rwanda and Burundi, 1889–1930: Chronology of a Slow Assassination," *International Journal of African Historical Studies* 18(1): 53–91; (2): 289–314.

Bruce, J. (2007) *Returnee Land Access: Lessons from Rwanda.* London: Humanitarian Policy Group, Overseas Development Institute.

Castells, M. (1996) *The Rise of the Network Society, The Information Age: Economy, Society and Culture, Vol. I.* Cambridge, MA/Oxford: Blackwell.

Castells, M. (2007) "Communication, power and counter-power in the network society," *International Journal of Communication* 1(1): 238–66.

Chrétien, J-P. (1995) *Les medias du génocide.* Paris: Karthala.

Cornet (1995). Famine noire et regards blancs: la famine Rwakayihura dans le Rwanda des années vingt in: R. Devish & F. de Boeck & D. Jonckers (ed.), *Alimentations, Traditions et Développements en Afrique Intertropicale.* Paris : L'Harmattan.

Cornet (2005), *Action sanitaire et contrôle social au Ruanda (1920–1940). Femmes, missions et politiques de santé.* Louvain-la-Neuve: Catholic University of Louvain.

Coupez, A. (1979) "Préface," in: P. Crépeau & S. Bizimana (ed.), *Proverbes du Rwanda.* Tervuren: Musée Royal de l'Afrique Centrale, pp. 1–2.

Crépeau, P. & S. Bizimana (1979) *Proverbes du Rwanda.* Tervuren: Musée Royal de l'Afrique Centrale.

de Lacger, L. (1961) [1939] *Ruanda.* Kabgayi: Imprimerie de Kabgayi.

de Lame, D. (1996) "Refugees in South Kivu, Zaire," *Relief and Rehabilitation Network Newsletter* 5: 9–12.

de Lame, D. (1997) "Le Rwanda et le vaste monde: Les liens du sang," in: S. Marysse & F. Reyntjens (eds.), *L'Afrique des Grands Lacs. Annuaire 1996–1997.* Paris: L'Harmattan, pp. 157–177.

de Lame, D. (2004) "Mighty Secrets, Popular Commensality and the Crisis of Transparency: Rwanda through the Looking Glass," *Canadian Journal of African Studies* 38(2): 279–317.

de Lame, D. (2005a) "Bridging the Gap between High Politics and Simple Tragedies," *African Studies Review* 48(3): 133–141.

de Lame, D. (2005b) [1996] *A Hill among a Thousand. Transformations and Ruptures in Rural Rwanda.* Madison/Tervuren: University of Wisconsin Press/Musée Royal de l'Afrique Centrale.

de Mahieu, W. (1985) *Qui a obstrué la cascade? Analyse sémantique du rituel de la circoncision chez les Komo du Zaïre.* Cambridge/Paris: Cambridge University Press/Maison des Sciences de l'Homme, pp. 381–412.

Des Forges, A. (1999) *Leave None to Tell the Story: Genocide in Rwanda.* New York: Human Rights Watch & FIDH.

Des Forges, A. (2006) "Land in Rwanda: Winnowing out the Chaff," in: F. Reyntjens & S. Marysse (eds.), *L'Afrique des Grands Lacs. Annuaire 2005–2006.* Antwerp/Paris: IOB/L'Harmattan, pp. 353–371.

Devisch, R. (1987) "Le symbolisme du corps entre l'indicible et le sacré," in: M. Jackson & I. Karp (eds.), *Personhood and Agency. The Experience of Self and Other in African Cultures.* Stockholm: Almqvist & Wiksell, pp. 115–133.

d'Hertefelt, M. & A. Coupez (1964) *La royauté sacrée de l'ancien Rwanda. Texte traduction et commentaire de son rituel.* Tervuren: Musée Royal de l'Afrique Centrale.

d'Hertefelt, M. & D. de Lame (1987) *Société, culture et histoire du Rwanda. Encyclopédie bibliographique1863–1980/87.* Tervuren: Musée Royal de l'Afrique Centrale.

Dowden, R. (2004) "The Rwandan Genocide: How the Press Missed the Story," *African Affairs* (103): 283–290.

Foucault, M. (1977) [1975] *Discipline and Punish: The Birth of the Prison.* New York: Random House.

Fujii, L. A. (2009) *Killing Neighbors. Webs of Violence in Rwanda.* London/New York: Cornell University Press.

Green, N. & L. Haddon (2009) *Mobile Communication. An Introduction to New Media.* Oxford/New York: Berg.

Guazzini (2001) "Riflessioni sulla identità di Guerra nel cyberspazio: il caso Eritreo-Etiopico," *Africa (Roma)* LVI (4): 532–572.

Guichaoua (1989) *Destins paysans et politiques agraires en Afrique centrale. L'ordre paysan des hautes terres centrales du Burundi et du Rwanda,* Paris: L'Harmattan.

Hannerz, U. (1992) *Cultural Complexity. Studies in the Social Organization of Meaning.* New York: Columbia University Press.

Hanssen, A. (1989) *Le désenchantement de la coopération. Enquête au pays des mille coopérants.* Paris: L'Harmattan.

Herman, E. S. & D. Peterson (2010) "Rwanda and the Democratic Republic of Congo in the Propaganda System," *Monthly Review Press* 62(1): http://monthlyreview.org/100501herman-peterson.php.

Higiro, J-M. V. (1992) "Rwanda: La voix de son maître," *Dialogue* 154: 29–37.

Hirsch, B., Y. Potin & M. Godelier (2005) "La parenté et l'histoire: entretien avec Maurice Godelier à propos des Métamorphoses de la Parenté," *Afrique et Histoire* 4: 247–281.

Horst, H. A. & D. Miller (2006) *The Cell Phone. An Anthropology of Communication.* Oxford/New York: Berg.

Ingelaere, B. (2010) "Do We Understand Life after Genocide? Center and Periphery in the Construction of Knowledge in Postgenocide Rwanda," *African Studies Review* 53(1): 41–59.

Kagabo, J, (1995) "Après le génocide. Notes de voyage," *Les Temps Modernes* 583: 102–125.

Kayihura, C. (1950) "Ceux qui ont bu au même chalumeau que Kabare," *Grands Lacs* 135: 37–42.

Lenoble-Bart, A. & Tudesq, A.J.(2008) *Connaître les medias d'Afrique subsaharienne.* Paris: IFAS/IFRA/MSHA/Karthala.

Liebhafaky Des Forges, A. (2011) [1972] "Defeat Is the Only Bad News. Rwanda under Musinga 1896–1931,"Madison: University of Wisconsin Press.

Linden (1977), *Church and Revolution in Rwanda.* Manchester: Manchester University Press.

Longman, T. (2010) *Christianity and Genocide in Rwanda.* New York: Cambridge University Press.

Malinowski, B. (1922) *The Argonauts of Western Pacific. An Account of Native Enterprise and Adventure in the Archipelagoes of Melanesian New Guinea.* New York: Dutton.

Maquet (1954) *Le système des relations sociales dans le Ruanda ancien.* Tervuren: Musée Royal de l'Afrique centrale.

Meyer, H. (1911) "Reiseberichte von Professor Dr Hans Meyer aus Deutsch-Ostafrika," *Mitteilungen aus den deutschen Schutzgebieten* 24: 219–221; 342–359.

Meyer, H. (1914) "Der Kagerafluss in Ostafrika und die Ruandabahn," *Koloniale Montsblätter* 1: 6–21.

Meyer, B. & P. Geschiere (1998) "Globalization and Identity; Dialectics of Flow and Closure," *Development and Change,* 20(4), 601–615.

Mushara, H. & C. Huggins (2004) "Land Reform, Land Scarcity and Post-conflict Reconstruction: A Case Study of Rwanda," *Eco-Conflicts* 3. Nairobi: African Centre for Technology Studies.

Newbury, C. (1988) *The Cohesion of Oppression: Clientship and Ethnicity in Rwanda 1860–1960.* New York: Columbia University Press.

Newbury, C. (1991) "The 'Rwakayihura' Famine of 1928–1929: A Nexus of Colonial Rule in Rwanda." *Histoire sociale de l'Afrique de l'Est.* Paris: Karthala, pp. 269–285.

Nsengiyumva, A. & A. P. Santiago (2007) "Rwanda: Balancing National Security and Development," in: E. J. Wilson III & K. R. Wong (eds.), *Negotiating the Net in Africa.* London/Boulder: Lynne Rienner, pp. 85–103.

Ntezimana, E. (1980) "Coutumes et traditions des royaumes hutu du Bukunzi et du Busozo," *Etudes Rwandaises* 13(2): 15–39.

Pottier, J. (2002) *Re-imagining Rwanda. Conflict, Survival and Disinformation in the Late Twentieth Century.* Cambridge: Cambridge University Press.

Pottier, J. (2006a) "Land Reform for Peace? Rwanda's 2005 Land Law in Context," *Journal of Agrarian Change* 6: 509–537.

Pottier, J. (2006b) "Roadblock Ethnography: Negotiating Humanitarian Access in Ituri, Eastern DR Congo, 1999–2004," *Africa (London)* 76(2): 151–179.

Prunier (1995) *The Rwanda Crisis 1959–1994.* London: Hurst & Co.

Reisdorff, I. (1952) "Enquêtes foncières au Ruanda," Unpublished report.

Reporters without Borders, 2011–2012. World Press Freedom Index 25.01.2012, //en.rsf.org/IMG/CLASSEMENT_2012/C_GENERAL_ANG.pdf, accessed on April, 12th 2012.

Republic of Rwanda (2010) *Rwanda Vision 2010*. Kigali.

Reyntjens, F. (1985) *Pouvoir et droit au Rwanda*. Tervuren: Musée Royal de l'Afrique Centrale.

Rukebesha, A. (1978) "La communication dans le Rwanda ancien," *Gazette de la Presse de Langue Française* (27–28): 14–16.

Rukebesha, A. (1985) *Esotérisme et communication sociale. Le cas du Rwanda*. Kigali: Printer Set.

Ruzibiza, A. J. (2005) *Rwanda. L'histoire secrète*. Paris: Panama.

Saur, L. (2004) *Le sabre, la machette et le goupillon. Des apparitions de Fatima au génocide rwandais*. Bierges: Mols.

Sirven, P. (1984) *La sous-urbanisation et les villes du Rwanda et du Burundi*. Thèse de Doctorat d'etat en geographie soutenue a l'Universite de Bordeaux, 1984.

Smith, P. (1997) "Les tambours du silence," *L'Homme* 143: 51–163.

Stanley (1890) *In Darkest Africa*, London: S. Low, Marston, Searle & Rivington.

Steering Committee for Emergency Assistance to Rwanda (1996) *The International Response to Conflict and Genocide. Lessons from the Rwanda Experience*. Copenhagen: DANIDA.

Strathern, M. (1996) "Cutting the Network," *Journal of the Royal Anthropological Society* 2(3): 517–535.

Taylor, C. (1988) "The Concept of Flow in Rwandan Popular Medicine," *Social Science and Medicine* 27(12): 1343–1348.

Tripp A. M., R. Lemarchand, A. Habimana, A. Songolo, C. Newbury & D. de Lame (2005) "ASR Focus. Commentaries on Marie Beatrice Umutesi's Book *Surviving the Slaughter*," *African Studies Review* 48(3): 89–143.

Umutesi, M-B. (2004) *Surviving the Slaughter. The Ordeal of a Rwandan Refugee in Zaïre*. Madison: University of Wisconsin Press.

Uvin, P. (1998) *Aiding Violence. The Development Enterprise in Rwanda*. Westhardford: Kumarian Press.

Vambe, M. T. (2008) "Autobiographical Representations of the Rwanda Genocide and Black Diasporic Identities in Africa," *African and Black Diaspora* 1(2): 185–200.

Vansina, J. (2002) *Antecedents to Modern Rwanda: The Nyiginya Kingdom*. Madison: Wisconsin University Press.

Vanwalle, R. (1982) "Aspecten van staatsvorming in West-Rwanda," *Africa-Tervuren* 28(3): 64–78.

CHAPTER 2

Research Practice in Connections: Travels and Methods

Mirjam de Bruijn and Inge Brinkman

Introduction

In the edited volume *Les discourses de voyages*, Romuald Fonkoua (1998: 5–10) introduces travel literature, with exploration traveling as the center of social analysis. He refers to the fascination surrounding travel and the relationship between traveling and discovery. This perspective is similar to that of travelers at the turn of the twentieth century who set off to discover the world, to experience things that were unknown and new to them. Fonkoua adds another perspective to the discoverer, namely, the writer-traveler (*le romancier*). It is, in the end, travel itself that forms the basis of the writing by the ethnographer and the novelist. Obviously, as Fonkoua points out, discoveries and travel writing are also constructed by people "in the other world" and in this sense travel is always about interaction. The construction of the world in those days was largely inspired by travel and journeys and we are now taking up this notion of "travel as discovery" and extrapolating it to our own constructions of the world.

The interpretative basis of ethnography was already changing by the beginning of the twentieth century, with the focus on itinerary and thus on space shifting toward a specific place, namely, "there" where the ethnographer "arrived." This meant that traveling was no longer the center of analysis but that "the field" as a concept became ever more important. As Clifford (1992)

This chapter was written in the framework of the Wotro program "Mobile Africa Revisited" (W 01.65.310.00). For more information, see http://mobileafricarevisited.wordpress.com/.

noted, no studies exist of the planes on which the ethnographer arrived in "the field" and the itinerary itself started to disappear from the narrative. Traveling no longer formed the basis on which one was to come to an understanding of the world. This more recent type of ethnography became an established genre, with its own set of methods and norms. It has also become a crucial source of information for colonial rule, and this has been well-researched.

Historical studies on Africa generally started later than ethnography. Initially there was a focus on travels of discovery and, with reasoning from a Eurocentric perspective, the emphasis was on European initiatives in "discovering" new horizons. Livingstone, Stanley, and others became famous within this Eurocentric perspective but, by the 1960s, attention in African history was shifting to African people and their past. This was a fortunate move and meant a shift toward a more static conceptualization of both methods and themes.

It is interesting that the earlier notion of "travel as discovery" retained its aura for many colonial officers. Colonial reports often take travel experiences as their point of departure and explain the cultures that are "discovered" from a travel perspective. Portuguese colonial officers in southeastern Angola, for example, largely based their understanding of the area that was under their rule on travel experiences (see de Almeida 1936). Travel journals became part of the knowledge of businesses or geographical societies in Europe (see Schestokant 2003) and such reports may have contributed to the image of Cameroon and southeast Angola as areas of long distances, difficulties, and loneliness, but also of the availability of important resources and their exploitation possibilities, which, for example, led to the plantation economies in Cameroon. This perspective also shows in the colonial fascination for overcoming and bridging distances and implanting new technologies in the realm of transport and communication. Roads, motorized vehicles, telephone lines, and the telegraph were all deemed wonderful contributions to "development" and while these new technologies were being introduced, imaginations and ideas about "native cultures" and "civilization" blended with the making of colonial reality.

A century later, mobility and space are back in vogue in ethnographic writing and are receiving ever more attention in historical writing on Africa. A spate of studies has appeared that take "the itinerary" as their focal point. The recent work of Urry (2007) stresses mobility, or the lack of it, as being far more crucial in most people's lives than the notion of "place." It is thus much more logical to study the (im)mobility of people and goods than of a certain "place." This new emphasis follows a long debate in anthropology about the concept of "the field." In their famous article, Gupta & Ferguson (1992) questioned "the field" as the core of the ethnographic encounter. In their view, ethnographers study culture by going to places: "the field." The question, however, is whether "the field" as a "sited place" is a fruitful notion to adopt in the study of "culture." Should understanding culture not be spatialized instead of sited (cf. Olwig & Hastrup 1997)?

As a way out of this dilemma between space and place, Appadurai (1991) proposed the notion of deterritorialized scapes. Much earlier, network analysis had been used as a way to understand society (and not so much culture) in sociology. The network analysis approach came to be connected to discussions on globalization, and "network societies" (Castells 2004) were seen in the light of global developments in the realm of digital Information and Communication Technologies (ICT). Network analysis has heavily influenced debates in anthropology and the study of space has shifted toward the concept of "nodal points," as used in network analysis to interpret the meeting of "the local" and "the global" (cf. van Binsbergen *et al.* 2006).

Migration has always been one of the major topics in ethnographic writing (Olwig & Hastrup 1997). In African historical studies, transport, telecommunications, and technologies in the realm of mobility are receiving ever more attention (Gewald, Luning & van Walraven 2010). Oddly, this has not immediately led to a questioning either of "the field" or of the situatedness of culture. Migration used to be regarded as an anomaly, a temporary condition from which people would return to a sedentary life and it was only in the late 1990s that this idea was criticized and the principles of an anthropology of mobility were laid (Salazar 2010). Migration is considered a normal condition of life in many cases (de Bruijn *et al.* 2001). The critique on settled life as the norm and the study of migration of people have led to an emphasis on transnational communities and processes of social relating.

This theoretical shift from place to space has entered mainstream anthropology and our understanding of "culture." But, methodologically speaking, how can we study such spaces of social relating? Is this "space" in the sense of "practiced place," as was suggested by de Certeau (1984: 117)? Does this new approach constitute a return to mobility and travel as central in our understanding of the world as in Fonkoua's analysis? This would imply that not only our own travel experiences but especially those of others matter for research. Some scholars have attempted multisited research as a way out: no longer did the "field" consist of one place but it was conceived of as being multiple, with research being undertaken in various places at the same time. In this way, a fuller interpretation of processes of globalization would hopefully be arrived at. As Feldman (2011: 397) argues, "It [Multi sited ethnography] emphasizes the experience of local immersion while acknowledging the local's disaggregation into the Global." Yet this will not suffice as a method as it still proposes "site" as the focus of attention. Vuorela

(2002) explained that the places at both ends of the journey are studied in Transnational Studies too: the "in-between" is left out. She argues for more emphasis on what she calls the "transnational habitus": the lived experiences in space, the histories during the itinerary. It was seen as one thing to point out that many societies are best understood as traveling cultures (Clifford 1992) but quite another to study such societies on their own terms.

Connections and linkages are central in this volume. Such linkages may be formed through communication, which implies travel, especially in societies that are depicted today as being transnational, mobile, and dispersed in space. This suggests not only physical travel but also the travel of ideas and discourses. Examples of such studies that emphasize the overcoming of distance as part of the research field itself are limited but increasing in number. The innovative ethnography by Amselle (2001) is an example of an ethnography on connections as he follows the itineraries of a ritual. Another example is Hofmeyr's (2004) book on Bunyan's *The Pilgrim's Progress* that deals with the travels and translation of a missionary book. Ling's study on the mobility of ideas and discourses has informed a study on climate change by de Wit (2011) in which she shows that the itinerary changes a discourse so that the situational moment of the discourse always remains an unforeseen practice. A recent study on migration between Africa and Europe concentrates on itineraries (Schapendonk 2011) and it is clear that new transport, information, and communication technologies are heavily influencing mobility patterns in the twenty-first century. This is not to say that mobility and its related actions of travel and communication are signs of "modernity": people have always moved and created connections between places just as technologies that facilitate such connections have played a role throughout history too (cf. Gitelman & Pingree 2003). The "modernity" interpretation fails to draw on the rich past of this linking in society and the historical links between new and old patterns of communication. Connections can be studied in archives as well.

For Africanist historians, the mobility of ordinary people in their daily lives became an issue in the 1970s and 1980s when social history was also becoming a noteworthy trend in African history with labor migration to the mines, towns and large farms being interpreted historically. The African diaspora, flight, and exile, and the links between urban and rural Africa have entered African historical studies, although only to a limited extent. It is clear that mobility as a theme in African history is growing and related themes, such as space and landscape (Luig & Von Oppen 1997; Howard & Shain 2005), are likewise gathering momentum.

As far as purely archival research is concerned, it is obvious that research activities were relatively static and stationary. Historians have tried to interpret mobility in the past by sitting and reading in a building where documents are stored. Although metaphorically traveling in time, no real movement is at stake in the case of archival study. Oral history is usually different: reading introductions to oral history books often shows that some physical travel was involved. Such travel is not the center of any methodological reflection, not even if it concerns literature in which historical method is an explicit item. Thus in Adenaike & Vansina's (1996) volume on historical fieldwork in Africa, the index mentions all kinds of subjects from fieldwork duration to the translation of culture and from oral traditions to the political situation, but no explicit reference is made to mobility and research. Relations between historical method and mobility have hitherto remained largely untouched.

The authors of this chapter, Inge Brinkman and Mirjam de Bruijn, are currently involved in a research program that has communication as the central element of the formation of communities. Communication includes forms of travel but other forms such as mobile telephony too. This chapter concentrates on travel, as this is still an important way for people to connect and communicate despite all the sophisticated technological possibilities that exist today. Travel is part of a society and informs social change. Understanding social change includes the traveling itself, being part of the connection. The two experiences presented below relate to ethnographical research both in the present and in the past, and how to make the itineraries and the travel experience part of the ethnographer's body of data and thus analysis. Travel should become a research method. The case studies presented are based on recently conducted research in Cameroon and Angola.

The first case study presents Inge Brinkman's attempts to understand the mobile community of Angolans living on both sides of its border with Namibia. The crux of the southeast Angolan case is that war raged in the region from 1966 to 2005. Refugees resident in neighboring countries hardly had any means of maintaining contact with relatives and friends still in Angola. After 1998 the war became less intensive and peace accords were signed in 2002. This profound change sharply influenced patterns of mobility and social interaction. As peace and new ICT such as the mobile phone arrived approximately at the same time, they are both associated with new possibilities for contact and communication. Following the routes of a family network from Windhoek to Rundu to Menongue and then to Luanda, this case explores the methodological and interpretative aspects of travel, communication, and social networks.

The case study in Cameroon relates to the history of a family and how its members developed their social cohesiveness despite being dispersed all over the world. The history of communication related to colonial interventions has been part of this research and it includes the histories of travel of families over generations. Our methodologies were based on archival research and on physically following historical itineraries. This case study discusses how these methods of documentation and travel informed our knowledge of the mobility patterns of these families. In recent years much mobility has been invested in the mobile phone and as such mobile phone research itself has become the central endeavor of anthropological fieldwork. Instead of being ruptures, travel and distance are the social glue of these communities and have become part of their identities (cf. Vertovec 2004). The experience of travel was essential to understanding the working of these communities. Indeed, the form this travel took for ourselves as researchers reflected the communities' hardships, enthusiasm, and (dis)connections.

Life on the Namibian-Angolan Border: Comfort and Hardship

Inge Brinkman carried out fieldwork in Rundu in the Kavango region of Namibia in 1996 and 1997. At the time, her focus was the nationalist war in southeastern Angola, from where her interlocutors all hailed. Mostly poor and illegal immigrants, none of them had a telephone and over time she lost contact with them. She did not know whether they were still living in Rundu or had moved to other regions, especially as peace had come to Angola in 2002. The chances were that they had gone to live there but she was particularly keen on seeing the two research assistants she had worked with and hoped to be able to follow the routes of travel she knew they usually took at the time.

On arriving in Windhoek, I (Inge) asked about transport to Rundu and was immediately discouraged by the owners of the bed and breakfast where I was staying from traveling by local transport as they deemed this to be dangerous and uncomfortable. However, the next morning I took a taxi to the bus stop near the Katutura Hospital and asked around for transport to Rundu. Everybody kindly assisted me in finding the right minibus, checking in, paying the right amount, and loading my luggage on board. The whole process was efficient and well-organized; we just had to wait a little while for the minibus to fill up. When describing travel, the focus is usually on the process of moving. Yet waiting is in many cases an integral part of travel. Even the journey itself can at least partly be described in terms of waiting, as people doze off or glance through a newspaper as the car or bus moves, waiting to arrive. Active agency is more with the car than with the travelers in it.

This is not to say that travelers are entirely passive. There may be long moments of silence as the passengers stare out of the window or sleep. But people may reach out of the car by receiving or making a phone call on their mobile phone. This is at the same time a moment of interaction in the car, as the caller might ask the driver to turn down the volume of the music or other people may overhear the phone conversation. There are usually also other moments of interaction in a car: people start a conversation, exchange newspapers, share food, and ask each other's names. There may be moments of intensified interaction between passengers. A stopover leads to increased interaction and communication between passengers, as does an unexpected event, such as a near accident. On the whole, contact on this journey remained superficial: the road was good, the car went fast, everything went smoothly, and we arrived in Rundu safely. From the stories I had been told and based on my own experiences, I knew that this journey could have been made in a similar fashion ten years earlier.

In Rundu I soon met both of my former research assistants and their families. I asked them about the changes in communication they had seen, the possibilities for renewing family ties, the newly opened transport facilities, and their mobility and visiting patterns. Living in one of the family houses, I saw this firsthand, observing the changes brought about by peace in Angola and the new transport and communication technologies. I decided to follow the family trails. Usually family history is static as it takes place in one particular geographic location. In this case, however, I followed the route of these family ties and combined family history with the relatively new concept of mobility research. The family in Rundu gave me letters and pictures to deliver to their relatives in Menongue in southeast Angola and early one morning I arrived at the bus stop in Rundu to wait for transport to Menongue.

Mobility research in a marginal region like southeast Angola is very different from Windhoek and Rundu where transport is relatively well-organized and the road between the two centers is tarmac. Firstly, the number of passengers traveling is generally lower so the time spent waiting is relatively high and there was not even anybody around who was going in the same direction. Only at 7:30 did a car arrive. The driver took my luggage and invited me to get in; we then dropped off his girlfriend, went shopping, and visited his sister. By the time we returned to the bus stop we had chatted quite a bit. I knew his name, where he was from, and how long he had been a driver and in this sense this time spent "waiting" for enough passengers to arrive was not waiting at all: the driver's errands had been taken care of and we had engaged in conversation. In such circumstances, traveling in anonymity is far less likely. While waiting for the minibus to fill up, the driver and the passengers talked about the purpose of their journeys and their family visits and already we were becoming a group of travels who shared a common destination. It was also tiring though and we finally left Rundu at 10:30. The first part of the journey was on a tarmac road and most of "us" fell asleep. After a few hours we stopped and the driver and his passengers began to trade, went to buy goods at a local store, and exchanged the latest news with their trading partners. This was an interesting moment that offered the chance to expand one's social networks: the passengers introduced their fellow travelers to their acquaintances and new connections could be made.

As we had left late and stopped for some time and it was a long way to the border, I started wondering about what time we would arrive in Menongue. When we finally reached the border, it turned out that the minibus was not continuing and we would now have to look for other means of transport. The driver saw two Afrikaans-speaking men from Grootfontein who he had met earlier and said: "You had better ask them whether you can sit in the *bakkie*. You whites always help each other. You will not even have to pay and you will go straight there. Maybe you will even be there at 19:00. The other passengers may only arrive at 3:00 or 4:00 in the morning." At the border, everything was rather confusing and nobody knew precisely what to do. These chaotic aspects on the one hand led to heightened interaction between the passengers, forcing an exchange about plans and possibilities. Yet added to this were the interventions of the border officials and some contact with other travelers and drivers. As a foreigner, I had to wait at a different counter from my Angolan cotravelers and the Namibians had to go to another desk. There were only a few cars and they seemed full. Everyone started looking for transport to Menongue on an individual basis. The "we" as a group had disappeared. Some of the travelers did link up again when they managed to secure a seat in another vehicle and all of them except me stuck to local public transport. One of the Afrikaans-speaking men had addressed me at the border post and as I did not fancy arriving in Menongue in the middle of the night without a place to stay, I decided to ask for a lift. Once across the border, they waited for me to fetch my luggage and I stepped out of the local itinerary. I could no longer speak to the other passengers and I no longer observed anything except the canvas of the *bakkie*. Sitting in the back I could not even talk to anyone. Did I disconnect? At the moment when I took the decision to travel this way, I did not care. We had not managed to contact family relatives in Menongue and so I was not sure about accommodation; the road was bumpy and my back was sore. When I arrived in Menongue about 20:00 I found a family I already knew from an earlier visit and the next day I went to find the relatives who I had taken messages and pictures for. I later learnt that the other travelers had indeed arrived around 4:00 in the morning.

On the one hand, I had left "the bridge" and followed a route that the other travelers could not afford to take. On the other, flexibility and taking opportunities as they present themselves is a characteristic of these journeys.

Although I had left the public means of transport and no longer followed the families' way of traveling, the Namibian men who allowed me to ride in the back of their vehicle were just as local as any other Namibian. I knew that the members of the family also occasionally used private transport from their work if they had the opportunity. Yet, it was obvious that the hardship and length of the journey played a role and the fact that I had already traveled from Europe to Namibia, from Windhoek to Rundu, and from Rundu to Menongue, constantly meeting new people, new food, new housing, new localities. To what limits can academics go when doing mobile research in marginal contexts? Should I have "stepped out" more frequently or earlier? Would it have been different if I had been an Angolan scholar? Or younger? I had learnt a lot during the journey. Some of these matters seemed trivial and self-evident but experiencing them had given me deeper understanding, in the same way that wisdom does not equal knowing a fact. The importance of the road surface, the strains of waiting, the trading activities en route, the way people started a conversation and turned into a group of travelers that could, however, be dissolved in a very short time span, the news that spread even after arriving in Menongue: these issues had become clearer and more vivid. Obviously it was a presentist impression, giving hardly any historical information. The journey helped me understand historical patterns of mobility in two ways. One was by similarity; I could imagine the ways in which news could spread, trading networks develop, and new contacts be established. The other was, by contrast, how much more difficult things would have been without tarmac roads and motorized transport, and if there had been no money or shops, as it used to be in the past.

As a researcher I had also asked questions and realized that this possibility for traveling had not existed prior to the peace accords, although other routes into Angola from Namibia had been possible since 1996 and Menongue could be reached by road. From conversations with family members I had already learnt a lot. After the peace accords, travel to Angola was still difficult but relatively secure.

> Then I also went to visit them in 2003. I knew them so I was not afraid of going to Angola. Before, going to Angola was like—wow—are you going to Angola? Will you come back alive? Are you going to reach your destination? But at the time we went, you were sure you were going to reach it.[1]

In 2003 it had taken two full days to reach Menongue as cars could not drive faster than 20 km an hour for long stretches but travel time has been reduced by half. On arrival, people tell their hosts about these changes and news about road conditions, travel times, hassles at the border, new trading possibilities, and difficulties in securing transport from the border to Angola spreads fast. These "arrival narratives," a genre with a considerable tradition, help people to situate themselves in the wider context of social networks, mobility and travel, and economic and political connections in the wider world.

Cameroon's Family Histories and the Researcher: Experiencing Linking Technology

In 2006 I (Mirjam) started historical-anthropological research in Anglophone Cameroon, curious to understand mobility in relation to possibilities to communicate and changes in the social landscape. One of the remarkable features of this part of Cameroon is the amount of traffic, despite the poor roads. Travel seems to be part of society and even of being. Buses, which are always fully booked, have a life of their own. One can trade and meet friends but most importantly travel connects and that is what people do all the time; they connect with the places and people that matter to them, for instance, attending funerals and cultural celebrations. When I tried to understand family ties in this world of connections I soon discovered that travel and being on the road is part of such a family network. The only way to discover family ties and their meanings was through the experience of travel (Figures 2.1 and 2.2).

Ethnographic research often depends on coincidence. One such coincidence during a research visit in January 2009 was the discovery of a diary that was in the possession of the Yenkong family and had been written by their father who died in 2003. The diary covered the period from the early 1970s until 2003. Abraham Yenkong, the son of a king (Fon) from the palace of Baaba, a small region on the Ndop plains west of Bamenda, had been a police officer who had trained at Scotland Yard in Nigeria in the 1950s. Different factors in his life meant that he was not eligible to follow his father, the Fon. The first of these was that his level of schooling had turned him into a person who was too well-educated for the fondom and the second was that—and probably more importantly—his mother had fled the palace in the 1940s to participate in church life (see Nkwi's chapter in this volume). Abraham Yenkong, a notable of the palace, thus became one of the traveling elites of Baaba. His work as the head of various police posts in western Cameroon turned him into someone who was part of the educated elite that built Cameroon following independence. The first two decades after independence were full of hope and expectations, and Yenkong was a "developer" in this period. When an economic crisis hit Cameroon at the end of the 1980s and in the early 1990s, Yenkong was already retired and had returned to Baaba, though he lived in Ndop. His subsequent political actions were

Figure 2.1 Windhoek, Namibia, minibuses ready to depart for Rundu, © Inge Brinkman, 2009.

Figure 2.2 Remembering communication: the old vehicle of the family at a border post, © Mirjam de Bruijn, 2009.

directed against the incumbent President Biya and he joined the opposition separatist movement based in Nigeria.

Here I briefly describe Abraham Yenkong and his family's life in the first two decades following independence when Yenkong married his four wives with whom he had ten children. In 2011 the family's children live spread around the world. There is only one son still living in Baaba, but his journey was one of failure in educational terms so his only resort has been to turn to tradition and he has become an expert in traditional medicine. Three of Yenkong's wives are also in Baaba. The other children are in Bamenda, Douala, the Netherlands, and the United States where they live with other family members. Being spread around the globe does not make them any less closely related and today the mobile phone and the Internet help maintain relations (cf. de Bruijn *et al.* 2010, de Bruijn 2010). In the past, even though distances may have been shorter, they were no hindrance to family life, with travel and letters being the most important linking technologies then.

The diary covers the time when Abraham Yenkong worked in different places in the South Western Province of Cameroon as a police officer and is full of travel, transport, and visitors. The travel was undertaken by him and his family, his wives lived in different places, his children were spread across the South West and North West Provinces, living with various family members and going to school in different places. Reading and realizing this life of travel made me restless. Was this travel and mobility to be seen as displacement, instability and considered or instead, as discussions with the children revealed, joyful and part of life. Travel was part of their lives, as was being connected. The family currently lives in a huge space, which does not seem to bother them, and their family ties are over distance but this does not mean that they are not relevant or important. On the contrary, this is how the family functions.

To understand the working and dynamics of the family's ties and its upward and downward social mobility, I had to experience their connectivity. This started with their past connectivity through travel. So I decided to travel the itineraries of the family as they developed in postcolonial Cameroon. I hoped to discover on this journey what made them connect and what their connections were about. Was the technology of connecting similar to a bridge, a telephone, or a road? What is the grammar of connecting?

Abraham Yenkong had died in 2003 so I could not travel with him. One of his younger sons, however, agreed to travel the itineraries he had made with his parents with me. We followed the historical trajectories of the family in June 2009 on two journeys that brought us to Oku on the Nigerian border, to Limbe, to Kumbo, and back to Bamenda.

However, before we embarked on this journey, we started our (re)discovery of the family itinerary with the family's photo archives. The pictures that we found in many parts of the house showed the different settings in which the family had lived. They had moved from one place to another every four years, which is why they had not lived in Bamenda for a large part of their lives. I turned these scattered pictures into a photo album that showed the chronology of the travels and places where the family had lived. This album then acted as an introduction to discussions about their history and would direct us to these places and their experiences there as a family. The pictures show the family's cars, the wealthy life they had, and the many family members, like grandmothers, aunts, cousins, and nephews, who joined the family for shorter or longer periods in their relatively luxurious houses. The pictures not only tell their own story but clearly relate the emotions and memories of the family members. These emotions were informative in understanding family ties and their significance for each individual despite the distances involved.

The stories and pictures also show the organization of this family in space. Yenkong's children traveled up and down between family in Baaba and Babungo (the mother's village) and the places where their father was living. Some of them spent their school years in a village in the North West Province where they lived with their grandmother. An aunt came to Limbe to take care of the children for a long period and others like nannies also helped in the household. This family had a remarkable number of children, cousins, and nephews but also far-flung family members who lived with the family for schooling. They have remained connected to the family and this has led to a web of family and acquaintances in the two provinces in western Cameroon. To keep these relations going and stay in contact over distance, travel is important. Did this mean that these itineraries are part of the family, just like their houses, estates, and family celebrations?

The next step in the discovery of the connections was the travel that I undertook with the son, and that we later discussed with Yenkong's other sons, daughter, and favorite wife. Yenkong had both his own and work cars (Land Cruisers) that were good for the poor road conditions and at one of the border stations. During our travel we saw that the wreck of one of Abraham Yenkong's cars lingered as a trophy at the police post at the border. The roads were still in bad shape when we traveled along them in the rainy season in June-July 2009 in our own Toyota Land Cruiser. The son was astonished that they had not improved and recognized the bridges and places where they used to pass and sometimes stop for a break. When we tried to go from Mamfe to the border region, the roads became impassable for cars and we could only go by motorbike. This brought back memories for him of his travels in the different seasons with his parents when he was a small boy and not yet in boarding school. These were faraway places in those days. Nevertheless, the son sketched an almost inviting picture of this time and travel was part of the romance of family life when the children came together in the father's house for holidays. Then they were able to live life as the children of a rich policeman with their stylish clothes and their hair in b-bop and drive around in their father's smart car. These were beautiful times that influenced their lives and meant that they would not be able to return to the village forever. This life with their father had turned them into elites as well. The condition of the roads was part of that life.

During our journey we were visiting places, meeting people from those days and searching for the places that were on the pictures we had found back in Baaba. The main buildings we visited were the houses where the family had lived, the schools and the church they attended, and the Cuba-style bars they used to visit. One of the son's main reactions was that not very much had changed in 25 years, except for the atmosphere in the bar. This "nothing has changed" has to be taken very literally. The furniture from 30 to 40 years ago was still in the offices and when we visited the office in the hospital where his mother (the favorite wife) had worked as head of the hospital, the furniture from those days was still there. The same was true at a place where his father had worked. The visit to the various bars where we had seen the pictures of the father and mother dancing, celebrating and drinking, revived other memories and he recalled the holidays they had when the brothers used the cars and were the "big" boys. Some things had changed, like the border town of Oku where his father's house had become dilapidated and the police post had been abandoned. Just like the furniture that had not changed since the family lived in those places, the poor state of Oku was also a sign of Cameroon's economic downturn since the 1990s when the country was plunged into crisis. The travel and the memories clarified how difficult this change must have been. The pictures show the family living a wealthy life in Oku, Limbe, Kumbo, and Ejumojock but these places seem not to have developed since. Indeed, the favorite wife who had worked alongside the father had never had a pension and remembered those days when they could afford luxuries. The travel made this more explicit than stories could ever have done.

As a last step in our research itinerary we made a new photo presentation on my laptop to include the pictures taken on the road of the estates and houses and discussed our itinerary with the sister, the favorite wife, and a cousin. They relived moments in the past and were happy to see that things were still there but they did not appear eager to return themselves. Their reactions revealed that travel is the core of the family existence, emotions, and being.

Discussion

This chapter has discussed two cases of research travel in which the researchers became part of a network of family connections. In the case of Namibia/Angola, all mobility and travel was evaluated in terms of the war, the ongoing legacy of the war, and the current changes in transport and communication possibilities. Since the war ended, these possibilities have shown vast changes: family visits and contacts were impossible during the war and it was only afterward that people could travel once again. In this sense, the Cameroonian case shows greater continuity. The discussions of the past itinerary showed a continuity with family relations in the present. Today's travel takes them to South Africa, Europe and the United States and in our discussions, these travels merged with travel in the past. The photo albums of the children in these geographically distant places helped to sketch the new itineraries as continuity of their grandfather's time.

The Cameroonian case had a much wider variety of sources than the Angolan case. Apart from interviews, the researcher had access to a diary and photo albums but no historical sources such as those that existed for the Namibian/Angolan case and information about the past only came through interviews and written source material not directly related to the history of the families involved. Given the region's long history of marginality, illiteracy, poverty, and lack of technological infrastructure, the lack of source material than other oral history amongst people from southeast Angola comes as no surprise.

Sources other than the research journey do matter as travel in itself can hardly form a solitary research method. Only through prior conversations and the consultation of source material can travel be meaningfully used to trace histories of mobility. Learning about the routes and connections within a family network makes it possible to experience these travels and in this way, travel functions as an auxiliary to other research methods.

In the Angolan case, the prior knowledge obtained through interviews and fieldwork ten years earlier enabled the researcher to situate the travel experiences in terms of both contrast and similarity. The Namibian context showed continuity to a large extent as the researcher experienced travels that the family members had been making for a considerable time. Travels into Angola were different in this respect and even though the war had only ended in 2002, conditions, speed, comfort, and travel possibilities had vastly changed.

A major difference between the two experiences is the materiality of the itineraries. In the Angolan case the emphasis was on the poor conditions of the road and the difficulties encountered when traveling, which reflect a history of war. In the Cameroonian case, roads were also bad and the itineraries were never easy but the family under study had a car and was able to travel in relative comfort. This materiality of the travel is important information. Traveling is also about settling and about the occupied spaces. In the Cameroonian case study, the pictures of the past made it easy to access the spaces and the transformations that had been undergone. The materiality of the travel and where they lived show the material well-being of the family in those days and their relatively high status. It is clear that they were among the elite in those days, an elite that was working toward progress in the newly born nation of Cameroon. The "discovery" in this travel experience of the nondifference between now and then in road conditions, in the furniture that filled the family's past offices, and the difference in the state of the family houses can also help in analyzing the effect of the economic downturn of Cameroon in the 1990s. In the Angolan case, travel could again become a full part of people's social repertoire while mobility was sharply reduced during the war years. This reflects an important change in society and in the social fabric that may lead us to draw conclusions on recent developments in Angola.

Should we revert to a situation in which fieldwork is seen as "exploration" and "discovery"? This question has to relate to the pasts of the people we are studying: how have they lived in travel, space, and settlement/place? The methods chosen should depend on the research context and must not be based on preconceived models. In our cases, following the history of family networks by modes of travel led to experiencing rather than noting down routes and travel. These experiences rendered our knowledge more practical than abstract and deepened our understanding of the hardships, opportunities, emotions, and contacts involved. These experiences at times were rough. In the Namibian/Angolan case, conditions at a certain stage became so challenging that the researcher temporarily stepped out of the public-transport sector to avoid the risks encountered when arriving in an unknown destination at night.

Our practice of fieldwork in connections by traveling made us part of the family network in a different way from just interviews and participating in daily life. By carrying pictures and letters between family members and sharing travel experiences, the researchers to some extent became part of the family network in movement. It also offered interesting insights into the extension of the networks and their nonsitedness. This in turn facilitated extending the research connections and sharply contributed to the snowball sampling effect (MacGaffey & Bazenguissa-Ganga 2000: 24–25).

We realize that this experience of travel with fellow travelers on present or past itineraries is a very personal experience. Nevertheless, the sharing of these experiences and the feelings that accompany them are informative and help the researcher know the social. This experience has been an important way of "practicing" the itinerary and discovering the travel as a "practiced place" (de Certeau 1984: 117). Did we then not also risk the "construction" of this social beyond "reality"? Did we overemphasize the memory of travel by traveling ourselves and memorizing travel with the help of pictures, as in the Cameroonian case? But is this not the problem with all fieldwork experiences? This chapter explained how we understood travel as a practiced place. Analysis of these experiences of travel and connecting as a grammar of connecting and social change was not its aim. In other words, we have tried to focus here on method rather than on research content.

Our experiences have certainly shown that it is worth making a "discovery," an exploration of travel, or more generally of connecting part of our fieldwork. Travel of space does not, however, go without place and settling so both the travel itinerary and the places of settling should be part of this discovery. The bridges to connect have changed with time and today it is therefore not only in travel that we discover connections but also in phone conversations and in virtual spaces on the Internet, that is, in social media. The discovery of these fluid spaces should become part of our field experiences just like travel in order to understand the "grammar of connecting" and social change.

Note

1. Conversation 9, with a woman born in Rundu in 1978, held on December 5, 2009, in Rundu.

References

Adenaike, C. K. & J. Vansina (eds.) (1996) *In Pursuit of History: Fieldwork in Africa.* Portsmouth & Oxford: Oxford University Press.

Amselle, J-L (2001) *Branchements, Anthropologie de l'universalité des cultures.* Paris: Flammarion.

Appadurai, A. (1991) "Global Ethnoscapes: Notes and Queries for a Transnational Anthropology," in: R. G. Fox (ed.) *Interventions: Anthropologies of the Present.* Santa Fe: School of American Research, pp. 191–210.

Castells, M. (ed.) (2004) *Network Society, A Cross Cultural Perspective.* Cheltenham: Edward Elgar.

Clifford, J. (1992) "Traveling Cultures," in: L. Grossberg, C, Nelson & P. Treichler (eds.) *Cultural Studies.* New York: Routledge, pp. 96–117.

de Almeida, J. (1936) [1910] *Sul de Angola, relatório de um govêrno de distrito (1908–1910).* Lisbon: Agência Geral do Ultramar.

de Bruijn, M. E. (2010) "Africa Connects: Mobile Communication and Social Change in the Margins of African Society. The Example of the Bamenda Grassfields, Cameroon," in: M. Fernández-Ardèvol & A. Ros Híjar (eds.), *Communication Technologies in Latin America and Africa: A Multidisciplinary Perspective.* Barcelona: IN3, pp. 167–191.

de Bruijn, M. E., F. Nyamnjoh & T. Angwafo (2010) "Mobile Interconnections: Reinterpreting Distance and Relating in the Cameroonian Grassfields," *Journal of African Media Studies* 2(3): 267–285.

de Bruijn, M. E., R. van Dijk & D. Foeken (eds.) (2001) *Mobile Africa, Changing Patterns of Movement in Africa and Beyond.* Leiden: Brill.

de Certeau, M. (1984) *The Practice of Everyday Life.* Berkeley: University of California Press.

de Wit, S. (2011) "Global Warming: An Ethnography of the Encounter of Global and Local Climate Change Discourses in the Bamenda Grassfields, Cameroon." MA Thesis, African Studies Centre, Leiden: ASC/LU.

Feldman, G. (2011) "If Ethnography is more than Participant-Observation, then Relations are more than Connections: The Case for Nonlocal Ethnography in a World of Apparatus," *Anthropological Theory* 11(4): 375–395.

Fonkoua, R. (ed.) (1998) *Les discours de voyages, Afriques-Antilles.* Paris: Karthala.

Gewald, J-B., S. Luning & K. van Walraven (eds.) (2010) *The Speed of Change. Motor Vehicles and People in Africa, 1890–2000.* Leiden: Brill.

Gitelman, L. & G. B. Pingree (eds.) (2003) *New Media, 1740–1915.* Cambridge MA: MIT Press.

Gupta, A. & J. Ferguson (1992) "Beyond 'Culture': Space, Identity, and the Politics of Difference," *Cultural Anthropology* 7(1): 6–23.

Hofmeyr, I. (2004) *The Portable Bunyan: A Transnational History of The Pilgrim's Progress.* Princeton & Oxford: Princeton University Press.

Howard, A. M. & R. M. Shain (eds.) (2005) *The Spatial Factor in African History: The Relationship of the Social, Material, and Perceptual.* Leiden: Brill.

Luig, U. & A. Von Oppen (1997) "Landscape in Africa: Process and Vision," *Paideuma* 43: 7–45.

MacGaffey, J. & R. Bazenguissa-Ganga (2000) *Congo-Paris, Transnational Traders on the Margins of the Law.* Oxford, Bloomington & Indianapolis: The International African Institute, James Currey & Indiana University Press.

Olwig, K. F. & K. Hastrup (eds.) (1997) *Siting Culture: The Shifting Anthropological Object.* New York: Routledge.

Salazar, N. B. (2010) " 'Towards an Anthropology of Cultural Mobilities," Crossings' *Journal of Migration and Culture* 1(1): 53–68.

Schapendonk, J. (2011) "Turbulent Trajectories, Sub-Saharan African Migrants Heading North." PhD. Thesis, Nijmegen University.

Schestokant, K. U. (2003) *German Women in Cameroon: Travelogues from Colonial Times.* New York: Peter Lang.

Urry, M. (2007) *Mobilities.* Cambridge: Polity Press.

van Binsbergen, W., R. van Dijk & J. B. Gewald (2006) "Introduction," in: W. van Binsbergen & R. van Dijk (eds.) *Situating Globality, African Agency in the Appropriation of Global Culture*. Leiden: Brill, pp. 3–54.

Vertovec, S. (2004) "Cheap Calls: The Social Glue of Migrant Transnationalism," *Global Networks* 4: 219–224.

Vuorela, U. (2002) "Transnational Families: Imagined and Real Communities," in: D. F. Bryceson & U. Vuorela (eds.), *The Transnational Family*. New European Frontiers and Global Networks. Oxford & New York: Berg, pp. 63–82.

CHAPTER 3

Patriarchy Turned Upside Down: The Flight of the Royal Women of Kom, Cameroon from 1920 to the 1960s

Walter Gam Nkwi

Introduction

The history of northwest Cameroon can be better understood by analyzing the increasing possibilities for connecting among ordinary people. Members of one specific society in this part of Cameroon, the people living in Kom, have developed a special attitude to new forms of connecting. Their story is one of appropriating new connecting technologies that were introduced in colonial times, namely, the church, schools, motor vehicles, and roads (de Bruijn *et al.* 2007; de Bruijn 2010). When these colonial technologies were introduced, people appropriated them in various but meaningful ways, domesticating them in the way they could best understand them. In Kom, this was known as *kfaang*, and for the Kom people, *kfaang* connotes newness, innovation, and novelty in thinking and action, and the material indicators and relationships that result from it (W. G. Nkwi 2011: 1). This chapter considers the appropriation of these linking technologies and their interaction with society and how this has led to social change. The technologies introduced by the colonial regimes can be compared to a linking technology, like a bridge, since they connect different worlds. This historical chapter discusses this interaction but also how these connections were given a life, and how and when they were objects that were integrated as change. How did these connections become a resource? The specific case study highlighted here has to do with gender transformations in Kom society. The explanation as to why, from 1920 onward, women fled the royal palace to escape the rigid patriarchal situation there seems to fit the connections model that this volume is developing. The story starts when a new phenomenon, namely, the Christian church, intruded into Kom society, bringing not only a new theology but also clothing and new ideas concerning love, gender, and schooling from another world. These were the carriers of the newness that the women encountered through the church and that led to their definitive new position in society.

According to Thomas (2002: 272), "The importance and significance of methodology lies in the fact that the issue of what is studied is intimately connected with the question of how it is studied." Consequently, a range of different methods and sources was used in the compilation of this chapter in the belief that historical sources contain a greater or lesser degree of subjective partiality. These sources included materials found in the archives in Cameroon and oral interviews conducted with those who were implicated in the issues under discussion and also with those who just witnessed the events. Some of the children of the runaway royal women were contacted too, although the information received from them was minimal.

One of the royal women who left the palace was Juliana (pseudonym), whose short biography helps us better appreciate how escape transformed her and the socio-scape. Her story illustrates how a woman could "completely change" through her connections with the church in Njinikom where she lives. She has four children, one boy and three girls: Veronica lives in Njinikom and is a retired midwife; Mary also lives there and is a retired boarding-school matron; Felix, the only boy, lives in Douala and graduated in civil engineering from the Technical University of Athens, Greece; and her youngest daughter is a nun at the convent.

Juliana was born in Wombong, Kom, around 1908 and was among the first Christians to be baptized in Nso. She was taken to the palace during the nascent years of Fon Ngam's rule (1912–1926) through a *nchinto* (a palace guard), but escaped from Laikom in the middle of the night when the guards were asleep and fled to the church compound in Njinikom. The women sometimes bribed the palace guards in order to be allowed to escape but she never had the means to do so. "Many of us escaped . . . and the Fon was very annoyed."

News about what was going on in Njinikom, like the arrival of a new church, had reached her through people from Kom who visited Fujua market near the palace and also because Biwa, the first Fon's wife, had also escaped to Njinikom. Juliana's decision to flee was because she heard that the "White Fathers were doing good things (including distributing new clothes to newly converted Christians at Njinikom) for women who were accepted into the church." According to her, there were also men who had returned from

Fernando Po who were generally kind and gentle. "I thought that it was better than the palace where the palace guards never allowed anybody to move freely. . . . I also thought that at Njinikom my children could attend school since a school was opened in Njinikom shortly after the church."

Juliana's profile as one of the pillars of connection of royal women to Njinikom could be considered unique. Her life raises certain questions: why, how, and when did the women escape? How did this affect their society and the one they created? And how did these women use their new connections to their advantage?

The chapter starts by situating Njinikom and its history, with particular attention being paid to the introduction of the church and Western Christianity. The second part then examines some of the factors that explain why the royal women fled from Laikom to the church compound in Njinikom, and the third part considers the consequences of such a move, demonstrating that it helped better not only them but also the whole of Cameroon.

Njinikom: A Hub of Connectivity

This section considers Njinikom, a village in Kom as a "junction box" that connected various technologies for social change. Juliana recounted how she and many other women escaped to Njinikom so it is important to situate it in ethnographic, geographical, and historical terms. Njinikom was founded in the nineteenth century as *Ijin-ni-kom* (literally, over the Kom country) and Laikom, the capital of Kom, actually meant "Kom country." Following the arrival of colonialism, the name was given a sharp spelling twist and since then Njinikom came into use. Colonial and postcolonial ethnographers, anthropologists, and historians adopted the name Njinikom. Njinikom is one of the 42 villages that make up the Kom Fondom and is situated in the middle of the Fondom. Another school of thought claims that "it was a valley of witches and wizards. The Fon had thought that the Whiteman and the witches would be better bed fellows" (de Vries 1998: 42–43). However, these ideas need further research.

Njinikom was first farmed by the inhabitants of Laikom. The landlords were William Fulmai and Jacob Ngweih and as the population increased and expanded into the area, it became gradually more domesticated.[1] When the Mill Hill missionaries arrived after the First World War, they were not allowed to settle at Fujua by Fon Ngam, who saw their teachings as an aberration to the cultural practices of the land (Ndi 2005). This led to the foundation of the first church in Njinikom in 1927 (W. G. Nkwi 2011) and its introduction provoked a chain of activity, with the escape of the royal wives from

Laikom being just one. It was a serious development though because of its implications for gender relations and transformations in the Fondom.[2]

St. Anthony's Primary school opened in 1928, and was to become the first standard six school in the Bamenda Grassfields in 1936.[3] In 1935 a school for catechists was opened in Njinikom (Booth 1973); a vernacular school, as the colonial administration came to call it because the language of instruction was in the Kom language, run by the colonial government, was also opened in 1924. In the same year a new and wider tarmac road was built linking Njinikom with Bamenda. A post office was also opened in 1955 and the first Kom, James Nsah Neng Ndai, was a first Kom to buy a vehicle that allowed people from Kom to reach other distant places faster and easier. In 1959, a primary school for girls, the St. Marie Gorretti School, was opened in Njinikom (W. G. Nkwi 2011). The arrival of these institutions shows how Njinikom not only has grown in prominence but has also become a hub of connectivity with colonial technologies. Many people have become connected because of wanting to appropriate these various technologies.

The Church and Its Messenger—Timneng—The Champion of "New Men"

The introduction of the church and Christianity in Kom predates the 1920s, going back to 1912 when the Catholic Fathers of the Sacred Heart of Jesus started their work in Fujua near Laikom. The missionaries had barely settled in when the First World War broke out in Europe, and later in Cameroon, which led to their expulsion from Cameroon and Kom. Everything they had started to construct, such as their mission compound, fell into ruins.[4] The war affected Kom indirectly. The German colonial administration demanded the Fon of Kom, Fon Ngam, provide soldiers to fight Germany's enemies (France, Great Britain, and Belgium) and this led to the recruitment of many people from Kom into the German army. Prominent amongst them was Michael Timneng who was to come back and challenge the traditional status quo.[5]

After the Germans were defeated, former soldiers and other Cameroonians were interned at Fernando Po. Many converted to Christianity and, once back in Kom, they were among the first to be baptized as Christians. Chilver (1963: 119) noted that as many as 900 people were baptized and 1,500 people received instruction from a catechist in the principles of the Christian religion with a view to baptism between 1916 and 1919 in Fernando Po.[6] Podevin, the Divisional Officer for Bamenda, reported late in 1916 that "the majority of German soldiers recruited from Bamenda area to fight the Great War were from Bikom and Bali. An ex-soldier from Kom reported at this office and informed me that he was present when more than 150 Bikom soldiers and carriers embarked at Bata with Hauptmann Adametz and left for Fernando Po. He also stated that when he left Bata captain von Sommerfeld was still there with about 100 Bikom soldiers and carriers."[7] When the Fon of Kom reported to the British Administration in 1917, he also revealed to the Divisional Officer that there were people from Kom in Fernando Po. Podevin later said: "The chief of Bikom revealed that there were 130 kom ex-German soldiers interned in Fernando Po and more than 180 carriers."[8]

The two reports and Chilver's research suggest that the number of Kom at Fernando Po was probably around 700. Rev. Fr. Baumeister, who had worked at Fujua shortly before the outbreak of the war, is said to have recognized Michael Timneng at Fernando Po and convinced him to become a believer and to swear on the day of his baptism that he would carry the faith and spread it when back home in Kom.[9] Timneng was happy with his new identity because he felt that he could use it as a counterweight to threaten the authority of the Fon with whom he had been at loggerheads since before he was sent to join the German army.[10]

Timneng kept his word and erected a small chapel at Wombong, a quarter in Njinikom, where daily prayers were conducted. He had also learnt to read and write, a magic which gave him added respect. As early as 1921, Njinikom was already grappling with astronomical numbers of catechumens waiting anxiously to be converted at a new church under Timneng's leadership as vicar. The ex-soldiers and Timneng stood up to and defied the traditional hierarchy of the Fon and his executive council, the *nkwifoyn*. Chilver (1963: 119) again maintained that "many of the repatriated men were enthusiasts, uncompromisingly contemptuous of traditional mores."

The resistance of traditional mores by the people of Timneng's category led to a protracted tug of war between them, the new converts, and the Fon. One of Timneng's earliest converts was Thecla Neng, who was born around 1908. She had to trek about 140 km to Shisong, Nso, with other Christians for doctrine classes and was finally baptized there. She told me that the war of words (*i-wong a-wo*) between the ex-soldiers (*ghu-gheili-iwong*) under Timneng and the Fon led to the humiliation of the Fon, who was commanded by the colonial administration that was acting under the auspices of the colonial officer to open the chapel that had been locked in Njinikom on his orders.[11] Timneng's mobility, both geographically and socially, and that of his colleagues did not only create a new social order but attempted to offset the previous one. Their mobility had instilled in them new ways of thinking and doing things. The novelty of what they had learnt meant that they could stand up against the traditional norms. The church in the hands of new converts became the governing structure and started to criticize the powers of the Fon, which he clearly did not like. Timneng and a host of returned ex-soldiers represented the cynosure that attracted the royal women to the church at Njinikom.

What is more relevant here is the fact that these returnees attracted women by the way they dressed, talked, and behaved and, above all, because they were from Fernada Po. They showed that the Fon was not as divine as he appeared and this led the royal wives to realize that they too could disobey the Fon and escape to an area with a liberal doctrine.[12] The person who championed this move in oral and written narratives was Timneng.

The Royal Women

Unlike the other women in the political and social setup of the Fondom, the royal women belong to the palace. According to Kom oral traditions, they constituted the privileged, and belonged to the Fon. Unlike other women who were usually wooed by suitors, the royal women were those considered to be beautiful and who should be reserved for the Fon. They were selected by the palace guards, the *nchintos,* and, once chosen, cam wood was sprinkled on their door post and a royal spear pinned in front of their house at night.[13] The girl's father had to take her to the palace with some palm wine (a white milky liquid tapped from a palm tree). The girls wore royal bangles around their necks and on their right wrist to indicate that no man should set eyes on them, let alone make love to them. This was not the best form of contracting marriage although some of the women loved the idea.[14] Gradually the Fon had as many wives as he wanted, maybe not for conjugal pleasure but for practical necessity and the pride of being polygamous.

According to Kom tradition, having many wives signified the continuity of the lineage in terms of procreation. The more wives one had, the more children he had and more wives and children brought wealth and prestige to his individual person. This was not specifically the case in Kom alone. Mbiti (1969: 139) noted, for example, that "polygamy also raises the social status of the family concern. It is instilled in the minds of African peoples that a big family earns its head great respect in the eyes of the community. Often it is the rich families that are made up of polygamous marriages." Only persons of influence were allowed to practice polygamy and the Fon of Kom was influential and saw having many women as a sign of prestige. The women were guarded and guided in their daily routines by palace guards, who had all been castrated. It was such attitudes that perhaps forced these women to leave, as life in the palace would not appear to have been easy. The women slept on bamboo beds and were nearly naked as clothes were a scarce commodity at the time. The church and its doctrine acted as a liberator of these women.

Time and Escape Roads

The story of Juliana shows that the timing and method of escape from the palace were carefully calculated. The time was key but causal observation would suggest that the structure of the palace was important too as it could facilitate the women's escape. The palace has five separate lodges, each of which is connected to the other by a gate (*ajei'ng*). The Fon's lodge is in the middle of the yard and close by it is the apartment of the *nkwifoyn*, which is out of bounds to all women, as is the palace guards' own lodge. The palace is located on a hillside and the only way the women could escape was via the northeast wing where there is a bridle path running through the forest down the hill and up to Njinikom. This was the route the women used to escape.

The middle of the night was the best time for the royal women to escape and they usually took their young children with them. On the day the royal wife was planning to escape she would cook for the palace guards and then stay awake until the first cock crowed, which was the signal that "half the world was dead." In a world without electricity or torch light, these women used a firebrand of elephant grass in places that were too dark.[15]

The Church and the Appeal of Its Theology

Christianity appealed to women and Njinikom was the center where many (royal) women were converted. They liked the Christian doctrine and the material culture, such as new clothing, that accompanied their encounters with the church.[16] During my fieldwork, many women acknowledged that their first clothes were given to them either by Rev. Fr. Leonard Jacobs or by Leo Onderwater who were both priests in Njinikom. One of the women for whom conversion and dress had a direct bearing was Helena Adiensa. She was born at Wombong in 1920 and could clearly remember her first dress, which was given to her by Rev. Fr. Leonard Jacobs early one morning after mass. The priest called her into the house and wrapped the clothes in black paper and told her that if she was asked who gave her them she should say that they came from the church.[17] These clothes were usually distributed in secret and only when the women had become catechumens or fully baptized Christians, or when they came for morning or daily mass. The giving of clothing was a strategy to convert people, especially women, to Christianity. The royal women were naked and saw the distribution and wearing of dresses as something new and attractive.

The Christian doctrine liberated the royal women from bondage. They numbered more than 500, although some, like the nun Rev. Sr. Loreto, claimed in *Times Magazine* that the number was closer to 700. She maintained that these women were living in inhuman conditions, having been coerced into the palace. The Christian doctrine ran counter to pre-colonial practice, for example, with its doctrine of "one man-one wife" or monogamy. It also taught the sacrament of matrimony, where one wife and one husband are bound forever in poverty or in riches, in sickness or in health till death do them part. Marriage did not only have a spiritual undertone but it also had a material dimension. The couple got married in new outfits: a suit for men and a wedding dress with a white veil for the woman. The Christian doctrine, coupled with its material indicators, acted as an incentive for women to escape from marital structures they found constricting. Ndi (2005: 54–56) noted that in Nso, another Fondom in the Bamenda Grasslands with almost the same Christian background, most of the royal women escaped and went to the mission.

The church also stressed the doctrine of one transcendental and omnipotent God. This God was all powerful and all seeing, and there was no other God above this God. The Ten Commandments were also evoked. The royal women had lived with the understanding that the Fon was their spiritual leader, which he was in effect since he performed quasi-spiritual functions and was seen as the only God in the eyes of the women. The Christian theologies of one God was thus more attractive to the women and also challenged their former view that the Fon was all powerful as they had thought.

The Nun's Activities as a Trigger for the Women's Escape

At celebrations to mark his 50 years of service in Africa, Bishop Rogan declared that "his main interest had been the spiritual welfare and material upliftment of women and girls" (Ndi 2005: 404). Such pronouncements prompted the Roman Catholic nun Rev. Sr. Loreto to write a hypothetical article in a postenlightened manner for their in-house magazine. Its main thesis was the poor treatment of royal women in the Bamenda Grassfields and she used the Fon of Kom as her case study. This later became infamous following its publication by the white South African writer Rebecca Reyer (1953) in a book with the carefully crafted title *The Fon and His Hundred Wives: The Fon Is the Ruler of Bikom, in the Cameroons: He Has 100 Wives*. Loreto's article hit the international press and was picked up by the United Nations, which had become the world's policeman after the Second World War. It subsequently sent a delegation to investigate the matter in 1948.[18] According to O'Neil (1991: 104), "After interviewing the royal women individually the delegation gave opportunity to any of the Fon's wives to leave if they wish to go, and forty seized the opportunity to do so. It was reported that the Fon was philosophical about the loss of almost half of his wives."

New Romantics: "Love" and "Sex"

Many royal women, in Kom, were sexually starved, since Kom mores saw these women more in economic than emotional terms. The women ploughed the Fon's farms and cooked for palace guests but the situation changed when the women protested by moving out. They were prompted to do so because the church and its pioneers, who had just returned from the First World War, were alluring. Colonial reports support this view and, writing about the returnees who accompanied the church, the DO for Bamenda, Hunt, claimed that the return of the Catholic Mission to Kom with its emotional appeal attracted many young women, the wives of the chiefs among them. Writing about the flight of royal women, he claimed that:

> In particular this has been the case with the chief of Bikom, a man between 60 and 70 with over a hundred wives of whom some are 20. Some of these, mostly young, have left him to attend the mission church and refused to return to him unless he gives them facilities for conversion, of which he will not hear. The result is a bitter estrangement between him and the Christian congregation, of whom some have harboured and more than harboured the runaways, so that he has practically cut off communication with the Njinikom quarter where the church is. Seduction of the wives of their people has also helped to set the chiefs of Banso and Kom against the mission.[19]

Being old and frail, the Fon could hardly have been expected to satisfy the sexual needs of his many young wives. In this respect the appeal of Christianity for such women was obvious. The returning migrants not only represented Christianity but also something deeper in their eyes, namely, a different romantic aspect, which the women had never had at the palace. The returnees and Christianity symbolized "romantic love," which was new and attractive to the royal women: they introduced a "love of newness" (*iikong-i-kfaang*). Cole & Thomas (2009: 4–10), writing about love in Africa between the wars, maintain that "we cannot understand sex or intimacy without understanding ideologies of emotional attachment . . . and that claims to love were also claims to modernity." This appeared to be linked to what was going on between the Fon's wives and the new arrivals in Njinikom.

The number of royal women who became pregnant was evidence of the levels of sexual activity at the palace. Colonial reports were filled with such information and pointed to the fact that many of the royal women gave birth in the mission compound. Following the 1923 League of Nations Report, the Assistant Divisional Officer for Bamenda, E. H. Hawkesworth, did not mince his words: "There are no less than 55 of the chief's runaway women in Njinikom, fourteen of whom had given birth to children by persons unknown and several more were pregnant. Many ex-German soldiers are blamed for that pregnancy. There also gathered a large percentage of the ex-German soldiers of Bikom (at Njinikom) who had been repatriated from Fernanda Po."[20]

Royal Women in Njinikom and the 1958 Revolt

When the royal women fled to Njinikom, accommodation was found for them near the church compound. Njinikom had become a Christian village and was cut off from the other non-Christian villages in Kom country. From the 1920s to the 1940s the number of royal women living in this quarter rose. One of the royal wives, Nathalia Nayah, died in 1924 and was buried in the Christian cemetery, according to Christian tradition, instead of in Laikom or Kikfuini where the royal tombs were. P. N. Nkwi (1976: 162) maintains that any attempts to have her body exhumed were greeted with a great deal of (Christian) protest because they did not want to resort to non-Christian ways. It was only in 1926 with the arrival of Fon Ndi, who legitimized Christianity, that the corpse was finally exhumed and buried at Kikfuini (P. N. Nkwi 1976: 163).

Njinikom saw significant changes between 1920 and 1958. A wide tarmac road heralded the arrival of the first motorized vehicles; a post office and a girls' school opened and the British colonial enterprise was fading out. The political demands for independence were rife with two main political parties in British Southern Cameroons firmly established in Kom: the Kamerun National Democratic Party (KNDP) and the Kamerun National Congress (KNC) (Ngoh 2001).

The teachings of the church appeared to brainwash women to a certain extent although this is difficult to prove. For instance, on July 16, 1958, the women from the Roman Catholic church addressed an open letter to Bishop Peter Rogan stating that "women's rights" in Kom had started on July 8, 1958 (O'Neil 1991: 105). This involved not only women fleeing from "oppression" and consequently breaking the bondage but was literally about the women emancipating themselves in their own minds. Like the seed growing silently, the gospel of liberation became firmly rooted in the minds of the women of Kom.

Fon Ndi (1926–1954) legitimized Christianity in Kom on the grounds that royal wives were to either return to Laikom or go back to their parents.

He died in 1954 without seeing this wish fulfilled and was succeeded by Fon Lo'oh Ndiforngu (1954–1966), but much to his concern, the women continued to leave the palace under his reign too. In a letter to the DO who was on tour in Kom, Fon Lo'oh complained that:

> Nearly all my wives have deserted me and scattered about everywhere in Kom against Kom Native Law and Custom relating the Fon's wives is that no Fon's wives is allowed to stay anywhere in Kom other than the Fon's compound. Nobody has any right to converse with the Fon's wives except their relatives. But now the Fon's wives go about the town contrary to the custom.[21]

The continued flight of the royal wives had far-reaching consequences for Kom and beyond, one of which was the Women's Revolt. In 1958, the power of the Fon was challenged by more than 6,000 women who demonstrated against a colonial policy that sought to change their farming methods—and the role of the royal women cannot be extricated from this revolt.

The Colonial Ordinances and the Women's Revolt

The Southern Cameroons Agricultural Law of 1955 was largely responsible for this revolt. On the other hand, the royal women who had congregated in Njinikom were the ones who made the revolt possible.[22] In 1956, the Wum Divisional Authority Soil Conservation Rules were enacted to reinforce the 1955 law that had eight articles stipulating farm size, methods of cultivation, and restrictions on farmland. In Article 2, for example, farms were to be divided by a six-foot-wide grass strip that followed the contour of the land (Kah 2011: 68). This was the one that directly affected Kom. Those who contravened these rules were liable to a fine of up to ten pounds or two months' imprisonment, or both (Ritzenthaler 1961; Nkwi 1985; Diduk 1989; Konde 1990; Shanklin 1990; Westerman 1992; W. G. Nkwi 2003, 2010: 24–25).

This ordinance, excellent as it may have seemed, failed to win over the women. Instead, in 1958, the Agricultural Department enforced the regulation without sensitizing the women to the merits of the new techniques. Fines were imposed on defaulters and corn, beans, and potatoes were uprooted in Anjin by the Agricultural Assistant, Joseph Ndikum. Vertical ridging had characterized the traditional Kom method of farming since the middle of the nineteenth century and, according to the Kom, this symbolized *Abun-a-wain*, the ridge of the child. To ask them to radically change their methods without educating them was enough to cause trouble. A meeting was held at Yindo Mbah's compound in Njinikom in July 1958 to explain the rationale behind the 1955 Agricultural Law. He explained that the law was enacted with the knowledge of the political parties, that is, the KNC and the KNDP, and insisted that it would not be repealed under any circumstances.

The law required that women in hilly areas construct ridges horizontally and not vertically, as had been the tradition. An agricultural assistant was posted to Kom but without the brief of actually educating the women that the reason behind the law was to check soil erosion in hilly areas, like Kom. The women revolted and their actions caused a severe rupture in the Fondom that extended to the colonial administration. The movement was known as *anlu*. It lasted for three years from 1958 to 1961, during which time the Fon was constantly called by name, which is considered a taboo. His executive arm, *nkwifoyn*, was undermined by the women and men who did not sympathize with the women were ostracized. The Fondom came in effect to be ruled by women.

The revolt was started in Njinikom mostly by Christian women converts, apparently escapees from Laikom. de Vries (1998) and Ndi (2005) have argued that Christianity influenced the way it started. It was impossible to completely separate the revolt from the missionaries' doctrine on the equality of people and the liberal ideas that were inculcated among Christians. The revolt's ringleaders were undoubtedly early Christian converts like Thecla Neng, Muana, Juliana Chiambong, Bi'wa Theresia, and Josephine Fuam. In addition, the revolt suggests more crucially that new notions of gender and women's rights had already reached Kom as they undermined all the previous patriarchal hierarchies in the Fondom.

When considering the church and the connections of the royal women, the gender aspect is striking. Not only was the power base of the Fon undermined but also the relationship between the sexes. The church opened up the way for a new interpretation of gender relations that diverged considerably from the traditional mores in Kom. Kom women, for example, assumed positions that had until then been exclusively the preserve of men. For instance, one of the women who coordinated the Women's Revolt, Muana, was given a seat in the Kom Native Court. This was the first time that women had had representation at court (W. G. Nkwi 2003: 156). Although Muana was not per se a royal escapee from Laikom, a good number of the women who had left the tutelage of the Fon represented an example that many Kom women emulated. Women thus saw another way in which they could cast off the old yoke of control and this was done through the *anlu* revolt.

The *anlu* also set in motion a set of changes that went beyond the Kom Fondom and affected the elections and the very future of British Southern Cameroon. This constitutes the subject of another research project that has already been done elsewhere (Ritzenthaler 1961; Konde 1990;

W. G. Nkwi 2003). Apart from the *anlu*, which was a direct consequence of the commotion that took place in the mission compound, it was ma Juliana whose connection to Njinikom had far more consequences in Njinikom and in Cameroon as a whole.

Conclusion

This chapter has taken Kom in Cameroon and the church as a linking technology, and considered the effect it had on women and the social transformations they went through as they were being connected. The church as a linking technology carried a lot of objects and ideas that literally transformed the women's lives. In the stories we came across, these included clothes, which were new for the women, and the theology of the church that offered new ideas about gender and about sex and love. But the main carriers of the connections were the women themselves. It was the women who had a reason to escape and appropriate the church in their own way: it was a daring act that would change their lives forever.

The women's flight shows that, in connections, new regimes develop and old ones can give way completely. The royal women demonstrated that a new regime came into being in Njinikom, even if it was not particularly visible. This chapter has also shown that, with the introduction of colonial technologies and their appropriation by people through connections, unforeseen consequences are bound to develop, as was illustrated by the Women's Revolt.

The women described here experienced profound changes that were hitherto unknown in their history. They underwent conversion to a Western-type religion, Roman Catholicism, with all its canons. At the same time, they experienced the material aspect of conversion through the clothes they were given to mark their conversion. It is hard to accept that the women were brainwashed but this emerged quite clearly in the *anlu* revolt. Revolt was an age-old tradition that was resurrected in 1958. Although the women's uprising was considerably different from those in the past in terms of novelty, it, however, shows continuity and how they had emancipated themselves from patriarchal subjugation. These royal women experienced drastic changes in their lives in Njinikom.

It should also be noted that some issues are still outstanding. How did these women escape from the palace? How and when did they arrive in Njinikom? What types of new social relations emerged? Who was actually responsible for their pregnancies? All these questions and more could be taken up in future research. This chapter has highlighted, however, how the royal women who fled the Fon's palace turned the patriarchal regime that they had hitherto been under upside down through their connections. It is hoped that this research will open up a new window for further interpretation of African history.

Notes

1. Interview with Batholomew Nkwain, Atuilah, Njinikom, August 14, 2010.
2. The Fondom here refers to the Kom geographical area, which is made up of 42 villages and is ruled by a paramount ruler known as the Fon, who is guided by an executive arm known as the *nkwifoyn*. Kom shares its eastern boundary with the kingdoms of Oku and Nso and its southern frontier with Kedjom Keku or Big Babanki and Ndop Plain. Bafut is on the western border while Bum and Mmen are found to the north.
3. Politics in the Bamenda Grasslands are dominated by and organized around the Fondoms ruled by Fons. These Fondoms mostly grew out of conquest and the politics of inclusion and exclusion as a result of war, which led to the subjection of weaker neighbors. They were dominated by political and social hierarchies based on kinship/kingship and lineages, on social and political status. Most studies have focused on the Fondoms and the establishment of political hegemony through social organizations. For more information, see Chilver & Kaberry (1967); Rowlands (1978) and Dillon (1990).
4. File No. Sd 1916/3, Report on Bikom mission, 1917 (NAB).
5. Jacqueline de Vries (1998), Paul Nchoji Nkwi (1976) and Anthony Mbuwe Ndi (2005) have already done incisive work on Timneng.
6. Chilver's figures may have been for the whole of Cameroon.
7. File Ba (1916)1 Confidential Report, No.1/1916, Bamenda Division by G. S. Podevin, District Officer (NAB).
8. File Ba (1917)1 Annual Reports, Bamenda Division by G. S. Podevin, District Officer (NAB).
9. Interview with Andiensa Helen, Njinikom, September 14, 2008. She was a close follower of Timneng and maintains that this story was shared with her by Timneng himself.
10. de Vries (1998: 37–39) showed the antagonism between the Fon and Timneng before he was sent to join the German army.
11. Interview with Thecla Neng, Njinikom, September 28, 2008. See also File No. Ad (1929/24) Bamenda Division: Kom Assessment Report by G. V. Evans 9 (NAB).
12. Christianity and sex is an area that has been generally ignored in research in Africa although it is closely linked to pregnancy and the early years of Christianity. The subject requires further research.
13. Interview with Anna Ayumchua, Njinikom, August 20, 2010. She witnessed how some of these women were taken to the palace.
14. Ibid.

15. This story was narrated by Susanna Njang, whose mother, Juliana Niih, was a royal wife. Her mother had escaped from the palace too. Interview in Buea, September 7, 2011.
16. Interview with Helen Andiensa, Wombong, Kom, September 30, 2009.
17. Ibid.
18. Cameroon became a UN Trust territory in 1946. UN missions were constantly visiting the territory to gather petitions and to prepare it for independence, as was stipulated in Article 76 (B).
19. See File No.Cb/1929/2, Annual Report Bamenda Division for 1929 (NAB).
20. File No. 4363, Roman Catholic Mission Njinikom: Memorandum of ADO, E. H. Hawkesworth to Divisional Officer, Bamenda, June 27, 1923.
21. For more on the problems confronting the establishment of Christianity in Kom, there are numerous files in the National Archives Buea. See, for instance, File Ba 1927/1 Cameroons Province League of Nations Annual Report for 1927; File Ba 1929/1 Cameroons Province Annual Report, 1928; Sd 1931/1 RC Mission (Njinikom), general Correspondence.
22. Laws of the Southern Cameroons 1954, 1955, and 1956 containing the Ordinances and Subsidiary Legislation of the Southern Cameroons (NAB).

References

Booth, B. F. (1973) *Mill Hill Fathers in West Cameroon: Education, Health and Development, 1884–1970.* Bethesda: International Scholars Publication.

Chilver, E. M. (1963) "Native Administration in West Central Cameroons, 1902–1954," in: K. Robinson & F. Madden (eds.), *Essays in Imperial Government.* Oxford: Basil Blackwell, pp. 100–108.

Chilver, E. M. & P. M. Kaberry (1967) *Traditional Bamenda: The Pre-colonial History and Ethnography of the Bamenda Grassfields.* Buea: Government Printers.

Cole, J. & L. M. Thomas (eds.) (2009) *Love in Africa.* Chicago & London: University of Chicago Press.

de Bruijn, M. (2010) "Africa Connects: Mobile Communication and Social Change in the Margins of African Society. The Example of the Bamenda Grassfields, Cameroon," in: Mireia Fernandez-Ardevol and Adela Ros Hijar (eds.) *Communication Technologies in Latin America and Africa: A Multidisciplinary Perspective.* Universitat Oberta Catalunya: Agecia Catalona de Cooperacio al Desenvolupament, pp. 165–191.

de Bruijn, M., R. van Dijk & J-B. Gewald (eds.) (2007) *Strength beyond Structure, Social and Historical Trajectories of Agency in Africa.* Leiden: Brill.

de Vries, J. (1998) "Catholic Mission, Colonial Government and Indigenous Response in Kom (Cameroon)," Research Report, Leiden: African Studies Centre.

Diduk, S. (1989) "Women's Agricultural Production and Political Action in the Cameroon Grassfields," *Africa* 59(3): 80–106.

Dillon, R. G. (1990) *Ranking and Resistance: A Pre-colonial Cameroonian Polity in Regional Perspective.* Stanford: University of California Press.

Kah, H. K. (2011) "Women's Resistance in Cameroon's Western Grassfields: The Power of Symbols, Organisation and Leadership, 1957–1961," *African Studies Quarterly* 12(3): 68–93.

Konde, E. (1990) "The Use of Women for the Empowerment of Men in African Nationalist Politics: The 1958 Anlu in Cameroon," Working Paper 147. Boston: African Studies Centre.

Mbiti, J. S. (1969) *African Religions and Philosophy*. London: Heinemann.

Ndi, A. M. (2005) *Mill Hill Missionaries in Southern West Cameroon 1922–1972: Prime Partners in Nation Building*. Nairobi: Paul's Publications Africa.

Ngoh, V. J. (2001) *Southern Cameroons, 1922–1961: A Constitutional History*. Aldershot: Ashgate.

Nkwi, P. N. (1976) *Traditional Government and Social Change: A Study of the Political Institutions among the Kom of the Cameroon Grassfields*. Fribourg: Fribourg University Press.

Nkwi, P. N. (1985) "Female Militancy in a Modern Context," in: J. C. Barbier (ed.), *Femmes du Cameroun: Meres Pacifiques Femmes Rebelles*. Paris: Karthala, pp. 181–193.

Nkwi, W. G. (2003) "The Political Activities of *Anlu* in the British Southern Cameroons Politics, 1958–1961: The Case of Kom Fondom," *Epasa moto: A Bilingual Journal of Arts, Letters and the Humanities* 1(6): 154–175.

Nkwi, W. G. (2010) *Voicing the Voiceless: Contributions to Filling Gaps in Cameroon History*. Bamenda: Langaa.

Nkwi, W. G. (2011) *Kfaang and Its Technologies: Towards a Social History of Mobility in Kom, Cameroon, 1928–1998*. Leiden: Ipskamp Drukkers.

O'Neil, R. J. (1991) *Mission to the British Cameroons*. London: Mission Books Service.

Reyer, R. (1953) *The Fon and His 100 Wives: The Fon Is the Ruler of Bikom in the Cameroons: He Has 100 Wives*. London: Victor Gollancz.

Ritzenthaler, R. (1961) "Anlu: A Women Uprising in British Cameroon," *African Studies* 19(3): 460–475.

Shanklin, E. (1990) *Anlu Remembered: Kom Woman Rebellion of 1958–1961*. Amsterdam: Kluwer Publishers.

Thomas, N. (2002) "History and Anthropology," in: A. Barnard & J. Spencer (eds.), *Encyclopaedia of Social and Cultural Anthropology*. London & New York: Routledge, pp. 272–295.

Westerman, V. (1992) *Women's disturbances. Der anlu-Aufstand bei den Kom (Kamerun) 1958–1960*. Hamburg: Zentrum Zur Afrikanischen Greschichte.

CHAPTER 4

Beyond the Last Frontier: Major Trollope and the Eastern Caprivi Zipfel

Jan-Bart Gewald

Beyond the Frontier

Even among the arbitrarily drawn borders of Africa, those of the Namibian Caprivi Strip are a striking anomaly, jutting 500 km into the African continent. Determined in the boardrooms of Europe as part of an exchange between the British and German Empires, the Caprivi Strip was designed to give the German Protectorate of South West Africa access to trade and traffic on the Zambezi River in an exchange drawn up according to maps and not necessarily the navigability of the rivers.[1] From the establishment of the German Protectorate through to the setting up of the independent Namibian state, the Caprivi Strip existed beyond the collective imagination of the Namibian state.

That the Caprivi Strip even exists is something of an embarrassment. It is an inconvenient appendage to the country as a whole, as can be gauged by the way in which successive Namibian administrations have sought to portray it on its own maps. German South West Africa's war map, the *Kriegskarte von Deutsch-Sudwestafrika*, that was published by Dietrich Reimer in Berlin in 1904 completely ignores the Caprivi.[2] To be fair, the German administration had more than enough on its plate between 1904 and 1908 with wars in central and southern Namibia to be overly concerned with the cartography of the Caprivi Strip. Nonetheless, the *Kriegskarte* set the standard for a large-scale map of the territory as a whole. Namibia's famous Farm Map, which has been printed in consecutive runs at a scale of 1:1,000,000 since 1921, portrays the

Caprivi not as an integral part of Namibia but as a separate entity at a scale of 1:2,000,000.[3] Maps published after Namibian independence in 1990 have not been changed and the latest one to be published by the Surveyor General in Windhoek in 2007 continues to show the marginal nature of the Caprivi by representing the territory as a separate entity, at half the scale of the rest of Namibia.

The marginality of the Caprivi in relation to the newly independent Republic of Namibia came to a head in August 1999 when armed members of the Caprivi Liberation Front (CLF) staged an insurrection and attempted to secede from the Republic.[4] Unfortunately for the secessionists, they failed to secure the military airport at Mpacha near Katima Mulilo and the Namibian government was able to fly in forces to suppress the secession. Subsequent military operations led Caprivians to understand that whatever their marginal status, they were answerable to the Namibian state. Units of the Namibian army and police and the Special Field Force (SFF) arrested, beat, and tortured hundreds of people.[5] And four years after the event, at least 130 people were still being held without trial, at least 11 had died in detention, and at least 5 were being held incommunicado at a detention center over 1,500 km away near Mariental in southern Namibia.[6]

Establishing and Maintaining an Administrative Presence

Although administrative presence was recognized on paper in 1890, it was not until 1909 that the German Empire actually established a presence in the Eastern Caprivi Zipfel.[7] Essentially an enormous triangle of land defined by the *Rio Cuando* (under its multiple names of Mashi, Linyanti, and Chobe) as it flows south and then northeast to its confluence with the Zambezi, the Eastern Caprivi Zipfel is an enormous area of land that consists of forested Kalahari sands and extensive marshes that are liable to annual flooding. Not surprisingly, the area is host to a wide variety of wildlife and the territory was contested by various powers that were anxious to control access to its bountiful hunting products. Also known as the Linyanti, the territory forms the basis of innumerable hunting and trading legends.[8]

While both the Bulozi and Batawana kingdoms contested access to the Linyanti in the second half of the nineteenth century, numerous European hunters were making names for themselves in the area (Tlou 1985; Morton 1996). Frederick Courtney Selous's companion French died in a hunting accident in the Linyanti in the 1880s and in 1906 Arnold Hodson (at that stage Assistant Resident Magistrate but later Governor of Sierra Leone) accompanied Resident Commissioner Sir Ralph Williams on what was basically a hunting trip to the Linyanti (Hodson 1912: 148–173). Hodson's group came upon a party led by Europeans that had been hunting there on territory essentially subject to Imperial Germany, although there was no official German presence in the area (*Ibid.*: 169–171). The hunting must have been particularly good as four years later, in 1910, Hodson traveled with the most senior British official in Southern Africa, High Commissioner Lord Selbourne, to the Linyanti and British officials again crossed over and hunted on what was German territory to all intents and purposes (*Ibid.*: 264–347).

The Eastern Caprivi was occupied in 1914 by Rhodesian troops guided by the cattle trader Harry Susman and it became "the first Allied occupation of enemy territory in the Great War" (Trollope 1956: 113). Administrative duties were performed by the British authorities in the Bechuanaland Protectorate, with the area falling under the District Commissioner Kasane. Throughout the 1920s and early 1930s, the area remained bereft of any form of colonial administration. As such, the territory formed a magnet for poachers and hunters anxious to make a living from selling dried meat, hides, and ivory. In 1939 the Native Affairs Department of the Union of South Africa assumed control of the Eastern Caprivi Zipfel and Major L. F. W. Trollope was appointed Magistrate and Native Commissioner of the area, which was then formally declared a "native reserve" in 1940 (*Ibid.*: 114).

Lord Beyond the Frontier: Lisle French Watts Trollope

Three years after the victory of the Nationalist Party in South Africa's elections, the policy of apartheid (known euphemistically in English as the "policy of separate development") was being enforced in all its barbarity across the territories subjected to the control of the Union of South Africa (Bunting 1969).[9] At the same time, in 1951, Lawrence Green published *Lords of the Last Frontier*, the "Story of South West Africa and Its People of all Races." A successful author of countless popular books on South Africa's history, Green was certainly not a supporter of the apartheid government but was a firm believer in the benefits brought to Africa by the *Pax Britannica*.[10] His *Lords of the Last Frontier*, which has become a much sought-after piece of Namibiana, lauded the activities of "Manning of Kaokoland," "Hahn of Ovamboland," and "Eedes of the Okavango."[11] These men, as Green dramatically noted, were alone in bringing the benefits of the British Empire to the territories: in the case of Hahn, "roughly two-thirds the area of Ireland" (Green 1961: 230). Hahn was "deeply sun-tanned, with dark, greying hair," and gave Green (1961: 232) "an instant impression of energy." Manning was "a man among men," a "short, slightly-built adventurer" with "an extremely virile body" (*Ibid.*: 75), and Eedes was "a fair-headed, clean-shaven giant, looking ten years less than his age," whose "lion-scarred back and arms" were observed by Green (1961: 246) "when he [Eedes] washed in the mornings." All receive ample coverage in *Lords of the Last Frontier* but no mention is made of Major L. F. W. Trollope, who served as Magistrate and Native Commissioner from 1939 to 1953 at Katima Mulilo in the Eastern Caprivi Zipfel. That Major Trollope's posting was beyond the last frontier is all the more evident when one realizes that the Eastern Caprivi Zipfel does not even exist on the maps in the *Lords of the Last Frontier*. It was not that Trollope was in any way lacking in distinctive physical attributes, although they may not necessarily have been those appreciated by Lawrence Green. The British naturalist Sir Frank Fraser Darling noted in his diary that Trollope was "a prodigiously fat man who is a bachelor and a very cultivated type" and that "he is a born gardener and obviously loves doing it."[12]

In contrast to most civil servants, Trollope did not pass into the oblivion of a forgotten past. Instead, he has had the dubious fortune of making it into the numerous travel guides that cater for tourists to Namibia today. The *Lonely Planet* has gone to the extent of claiming that "the rather idiosyncratic magistrate Major Lyle [sic] French W. Trollope was posted to Katima Mulilo . . . long enough to be regarded as local royalty" (Hardy & Firestone 2009: 200). This was possibly due to the fact that Trollope had a fully functioning flush toilet installed in a hollow baobab that stood in the grounds of his residence in Katima Mulilo.[13] Apart from leaving a toilet in tree for posterity, Trollope also left a reputation that, although it exceeds the archival records, does show him for what he was, namely, a remarkable character. A specialist scientific advisor to the Namibian government noted approvingly of Trollope in a case brought before the International Court of Justice (ICJ) in The Hague:

> Eventually Trollope fell out with the newly elected National Party government in South Africa. Correspondence on file in the archives shows his increasing intransigence, which eventually resulted in an instruction to return to Pretoria. He ignored the order and when his replacement arrived, Trollope asked him if he had a permit to enter the Caprivi!
>
> (Alexander 1999: 323)

Trollope died shortly afterward and his grave, which is another of Katima Mulilo's tourist attractions, lies in "the small cemetery just outside the [Zambezi] hotel gate, with its large headstone for the previous magistrate of the Caprivi LEF [sic] Trollope" (Santcross, Ballard & Baker 2009: 171).

On a more serious note, prior to his posting to the Caprivi, Trollope had been the Assistant Native Commissioner in Windhoek, replacing Major Cocky Hahn while he was on leave, and had been in contact with Southern Rhodesia in 1936.[14] Trollope was appointed Native Commissioner and

Magistrate to the Eastern Caprivi Zipfel in 1939. This appointment, which even made it into North American press, was made with the explicit intention that Trollope undertake action against witchcraft.[15] An official statement even noted that Trollope's duties were to:

> Administer the territory, making full use where possible of native institutions, to combat the evil of witchcraft and to advise and assist in the gradual uplift of the natives.
>
> (*Ibid.*)

Appointed specifically to try cases of witchcraft, Trollope got off to a good start. The very first court case he dealt with as Magistrate in Katima Mulilo in November 1939 was one involving witchcraft.[16] And although he oversaw numerous other cases between 1939 and 1953, this one case at the very beginning of his career was to be his only witchcraft case for as long as he served as Native Commissioner and Magistrate in the Eastern Caprivi Zipfel. The bulk of his cases dealt with soldiers from the South African Native Labour contingent going absent without leave, being drunk, sleeping on duty, or losing all manner of army equipment.[17]

Instead of prosecuting people for witchcraft, Trollope went out of his way to get to know the territory that had been placed under his jurisdiction. In his first year, he wrote a report that nearly 50 years later would come to form part of the evidence used by the Republic of Namibia in its legal wrangling with Botswana at the International Court of Justice in The Hague.[18] In investigating the frontiers of his fiefdom, Trollope came into close and sustained contact with his counterparts in the Bechuanaland Protectorate, the District Commissioners for Kasane and Maun, as well as the District Commissioner Sesheke in Northern Rhodesia. The Trollope Redman Report completed by Trollope and D. C. Maun in 1948 determined the ICJ's ruling in favor of Botswana, whilst also emphasizing the close working relationship Trollope had with his counterparts in the Bechuanaland Protectorate (Perry 2000: 83).

It is in correspondence with his colleagues that one clearly sees Trollope's close links with his fellow officers in the immediate vicinity of Linyanti. In itself, this is not surprising given the distances involved but what is surprising is the fact that in effect Trollope maintained closer links with colonial officers in three separate and distinct colonial territories than he had with his formal employers, let alone the administrators of the mandated territory of South West Africa, of which the Caprivi Zipfel was officially an integral part.

By force of circumstance, all post to Trollope in Katima Mulilo was delivered to the post office in Sesheke, Northern Rhodesia, on the opposite bank of the Zambezi River. It was, however, a conscious choice on Trollope's part to negate any reference to the mandated territory of South West Africa or to the Union of South Africa in any of his formal correspondence. The date stamp used by Trollope as Native Commissioner noted his address as "Native Commissioner; Eastern Caprivi Zipfel; Katima Mulilo; P. O. Sesheke; Via Livingstone Northern Rhodesia." It is understandable that colleagues, friends, and acquaintances writing to Trollope should ignore any reference to South West Africa or South Africa, but the same cannot be said for Trollope. In addition, he provides no indication whatsoever of being part of the formal jurisdiction of the Union of South Africa in the Mandated Territory of South West Africa.

In this, Trollope stood in stark contrast to the administrators appointed by the apartheid regime after the Nationalist Party's victory in elections in South Africa in 1948. In contrast to these administrators, who were subject to a regime that was riven with modernization directly controlled and centrally led from Pretoria, Trollope was able to choose to do very much as he wanted in Caprivi.[19] Formally appointed as Native Commissioner in charge of the Native Reserve of Eastern Caprivi Zipfel, Trollope could simply ignore directives that were not to his liking and instead indulge in his own idiosyncrasies. One such indulgence was the building of his famous "laver tree" (see above) when Trollope was able to enlist the services of a work crew led by Chris Conradie of the Public Works Department in Northern Rhodesia.[20] Exactly how and why the Public Works Department of Northern Rhodesia came to be deployed in the Eastern Caprivi, which was formally a different territory and one not subject to the jurisdiction of Northern Rhodesia, is not to be found in the archives but the case illustrates the manner in which Trollope and his fiefdom functioned within the ambit of Northern Rhodesia but not South Africa or South West Africa.

Stationed in Katima Mulilo, Trollope showed little interest in the formal administration of taxation and labor censuses among the population of the Eastern Caprivi. Instead, he maintained an intense stream of correspondence with (natural history) museums, anthropologists and scientific societies around the world. Trollope's interests ranged widely, from the taxonomy of flora and fauna and the perceived ethnographic details of various tribes to the mythical "lost city of the Kalahari." The extent to which the Eastern Caprivi came to be seen by his correspondents as Trollope's personal fiefdom can be gleaned from such correspondence. When writing to Trollope in 1950, the Keeper of Antiquities at the National Museum of Southern Rhodesia in Bulawayo thus noted:

> Your information clears the matter so far as your own—Caprivi Strip—Subia are concerned and I have asked Mr. Desmond Clark to make further enquiries about the remnants of this tribe.[21]

Similarly, the District Commissioner in Maun in the Bechuanaland Protectorate, R. A. R. Bent, wrote to Trollope:

> I have had news of you frequently, and gather that you are flourishing in your beloved Caprivi, having spumed [sic] the flesh pots elsewhere.[22]

In other correspondence with Trollope, Desmond Clark, the Archaeologist and Curator of the Rhodes-Livingstone Museum in Livingstone, Northern Rhodesia, addressed Trollope as though he were in Northern Rhodesia.[23] Trollope's detailed knowledge of western Northern Rhodesia comes to the fore in his correspondence with Tim Brent, the Veterinary Officer in Mongu, Barotseland. In a letter, Trollope dealt with and questioned some of the particulars of the Carp Expedition that had passed through Barotseland, noting that he, Trollope, had visited practically all of the villages north and south of Singalamwe, that is, all the villages along the Zambezi in western Barotseland.[24] In his correspondence with all and sundry, Trollope listed his address as being in Northern Rhodesia and clearly felt himself to be outside the Union of South Africa. In a letter to Bernard Carp when referring to fauna, he noted "it does not resemble very much the bushpig of the Union," explicitly placing himself beyond the confines of the Union of South Africa.[25]

The Fall

P. S. Was extremely sorry to hear about your recent very nearly expiral, but very glad indeed to hear you are alright.[26]

Major Trollope was a bachelor and, as a newspaper article put it, "he never married but was assisted in entertaining his many guests by his sister, Alice."[27] In the course of his stay in Katima Mulilo, Trollope became acquainted with a young man named Anderson Luendo, who he provided with clothing and whose school fees he also paid. Luendo, however, turned out to be a bit of a dark horse and was arrested and convicted of burglary in Sesheke, Northern Rhodesia, on the northern bank of the Zambezi River directly opposite Katima Mulilo.[28] In 1952, Luendo, who was then 19, was arrested by Trollope and placed in the magisterial lockup in Katima Mulilo. Exactly why Luendo was arrested is not evident from the subsequent court case and he was never formally charged following his initial arrest.

In Luendo's own words, he was confined to prison for "about three months without trial, and was given the job of looking after cattle which did not belong to the Government but to the magistrate Major Trollope."[29] Luendo's subsequent statements, as well as those of the magistrate who came to replace Trollope, A. B. Colenbrander, indicate that detention in the Katima Mulilo lockup depended to a large extent on the cooperation of the person arrested.[30] In his statement, Luendo noted that whilst in prison "I used to go to work about 6 a.m. and I came back about 1 p.m. The policeman then shut me up in the cell until 4 p.m. when I had to go back to work. I ate at about 5 p.m. Sometimes he opened the cell and got me to cook for him."[31] Luendo stated that "the reason why I escaped from the Caprivi prison was because I did not get on well with the policeman who was in charge of me." And he added that:

> It was because of this long waiting without trial that I felt angry. So, one Sunday night I took an axe. I took the axe not because I intended to kill Major Trollope but because I wanted to make him feel pain because he had made me feel pain.[32]

On the night of October 27, 1952, Luendo attacked Trollope with an axe. At the trial, Trollope recollected what happened on that fateful night. Sharing a house with his sister, Trollope slept on the enclosed verandah on the east side of the house while his sister slept in a room on the west side. Trollope's bed was near to the famous Baobab water closet in which a paraffin lamp had been placed as a night light. Trollope recounted that he had gone to bed at "the normal time," had read for about 1 to 1½ hours, put out the light above his head and fallen asleep "almost immediately."[33] In the words of Trollope:

> My next recollection is wakening up during the night by feeling a violent pain in my right side . . . The pain was considerable and my senses confused but my first clear recollection is that I found myself standing next to my bed with the axe Exhibit I in my hands.[34]

Trollope survived the attack, and the following morning a fully loaded revolver belonging to the prison guard was found lying outside the door next to Trollope's bed. Trollope was confined to bed for two days and was in considerable pain for weeks afterward. A few days after the attack, Trollope saw the accused wearing clothes "which belong to me and are some of those I issue to my house boys."[35]

Anderson Luendo was found guilty and sentenced to nine months in prison with hard labor and eight lashes.[36] Trollope never recovered from the attack and died four years later while undergoing medical treatment in Bulawayo.

Hard Borders

The attack on Trollope heralded the end of his reign in Eastern Caprivi. In its aftermath, Trollope was replaced by a new magistrate, A. B. Colenbrander, who, in contrast to Trollope, was prepared to vigorously pursue an anti-witchcraft campaign in the area. In the years that followed, the apartheid regime adopted a program whereby the Eastern Caprivi was separated from the surrounding territories of Bechuanaland and Northern and Southern Rhodesia, and postal services were rerouted via Windhoek. In 1958 the Native Commissioner wrote to the Secretary for Native Affairs in Pretoria suggesting that military aircraft patrolling the borders of the Union of South Africa deploy a detachment of five or six instructors from the South African army:

> [to] provide a small scale display of the use of machine guns and mortars under conditions of war.[37]

According to the Native Commissioner, this would have a "healthy impact on the psychological condition of the natives on the other side of the border."[38] Whereas previously the District Commissioner in Sesheke on the northern bank of the Zambezi had maintained contact with the Magistrate in Katima Mulilo on the opposite bank and the two had visited one another freely, the Magistrate in Katima Mulilo now had to inform the District Commissioner for Sesheke:

> In so far as granting permission for what I presume will be a party to visit the area concerned I regret that I have no authority to do so but feel certain that if an application is submitted from your side to the secretary for Bantu Administration and development, P.O. Box 384, Pretoria, he will grant you the necessary authority.[39]

Less than ten years after the attack on Trollope, the Eastern Caprivi had effectively and deliberately been cut off from Sesheke. Zambia's independence in 1964 and the subsequent Bush War, which would finally end in 1990, ensured that what had been a soft border became a hard border. Any association by administrators subject to the wishes of Pretoria with administrators subject to the wishes of Lusaka were vigorously discouraged.

Conclusion

Formally in the employ of the administration of the mandated territory of South West Africa, Trollope, who had previously been Assistant Native

Commissioner in Windhoek, served as Magistrate and Native Commissioner beyond the "last frontier" for 14 years. Although nominally in the employ of an administration that was presided over by a government that sat in Cape Town and Pretoria, Trollope had in practice become lord of his own domain by the end of his tenure in Katima Mulilo and was far more integrated into the administrations of Northern Rhodesia and Botswana than the Union of South Africa. The connection of South Africa, at the furthest fringes of its rule, was mediated through the person of Major Trollope, who was at liberty to impart to South African rule his own personal insights, insights that did not necessarily bear any resemblance to the formal strictures of administration as determined in Pretoria and Windhoek. The coming to power of the Nationalist Party of South Africa, through heralding the end of Trollope's rule, effectively broke the connection that existed between South Africa and its neighbors as mediated through Trollope, and brought about far-reaching changes in the Eastern Caprivi. No longer was it to be administered by the whims of a single person. The territory was instead to be consciously freed from the surrounding territories and subjected to direct rule enforced from Pretoria. Soft borders, which had allowed the free movement of administrators and people back and forth across the Zambezi, came to be replaced by a border that was enforced by military might from 1966 until Namibia's independence in 1990. In effect a new connection, now determined by military considerations and the perceived security concerns of South Africa's apartheid government, came to determine its relations with countries on the fringes of its empire.

Notes

1. For a detailed introduction to the Eastern Caprivi, and Katima Mulilo in particular see Zeller (2009). For an overview of the German and British administration prior to 1956 see Trollope (1956) and for a description of the Caprivi in the heyday of apartheid, see Logan (1972).
2. Originally published as *Kriegskarte von Deutsch-Sudwestafrika* 1:800,000, Berlin: Dietrich Reimer Verlag 1904. Republished in 1994 by the National Archives of Namibia.
3. On the history of the Farm Map, see Miescher (2007).
4. The CLF refers to Trollope in its propaganda, seeking to justify its specific status separate from the nation-state of Namibia.
5. The SFF comes under the 1990 Police Act. However, unlike the regular police, no educational requirements are necessary for SFF membership and it has come to be staffed primarily by PLAN (People's Liberation Army of Namibia) veterans and political appointees. During operations in the Caprivi, they reported directly to the head of state and not to the Minister of Home Affairs, thus removing any parliamentary supervision.

6. For details of the insurrection, see Amnesty International (2003).
7. For more details on the intricacies of the German presence in the Eastern Caprivi, see Fisch (1996).
8. Macmillan (2005: 102) notes how the establishment of German and British rule "complicated the Susman brothers" trading relations with people in the two territories where trade had been unregulated.
9. The complete text of Bunting's work is accessible at http://www.anc.org.za/books/reich.html.
10. Green (1961: 17) noted, "Peace came to Africa, thanks to the efforts of various powers. I am thinking especially of the Pax Britannica in tropical Africa."
11. The positive sentiments of Lawrence Green regarding these men are not shared by all. Elsewhere, Hayes (1996) has torn "Cocky Hahn" from his pedestal.
12. For his diary entry for September 28, 1956, see Boyd (1992: 95).
13. During the guerrilla war that developed in the 1970s and 1980s, South African forces often had themselves photographed on this toilet. Today the Internet is littered with photographs of many of these South African veterans "ou Manne" revisiting the haunts of their past in 4 × 4 convoys. One itinerary, which includes visits to 32 Bn graves and Fort Doppies, notes the GPS coordinates for the tree: "We have a peek at the toilet which has been built into a large Baobab tree at S17 29.329 E24 16.693." http://www.4x4community.co.za/forum/archive/index.php/t-13543.html, accessed April 14, 2009.
14. NAN, LGO Gobabis Magistrate, Aminuis Native Reserve, Precedents File, 2/10/2, L. Trollope, Assistant Native Commissioner, Windhoek, June 13, 1936, to Mr. Geard. & NAN, LKM L. Trollope in Katima Mulilo, December 28, 1952, to Bernard Carp.
15. "Pretoria Will Undertake War against Witchcraft," *The Lima Times*, January 11, 1940, p. 14. & "Pretoria to Undertake Drive on Witchcraft," *The Post-Democrat* 20(34): January 19, 1940.
16. National Archives of Namibia, Windhoek (NAN), Magistrate Katima Mulilo (LKM) 1/1/1 Criminal Cases, Record of Proceedings 1939–1978.
17. NAN, LKM 1/2/1 Criminal Record Book, 1939–1957, January.
18. International Court of Justice, "Case Concerning Kasikili/Sedudu Island (Botswana/Namibia)," Memorial of the Republic of Namibia, February 28, 1997, pp. 202–203. Appendix 58 L. Trollope, Report on the Administration of the Eastern Caprivi Zipfel, 1940.
19. Following the Nationalist victory, all administrators were issued with the same orders and expected to enforce the same policies, irrespective of where they were in the Union of South Africa and South West Africa. For instance, Trollope's successor received reams of documents and policy guidelines relating to "influx control" and the employment of domestic servants that had no bearing on the rural setting of Katima Mulilo on the fringes of South Africa's jurisdiction.
20. The Chiel, "On the Chain Gang," *Daily Dispatch*, Friday January 21, 2000, http://www.dispatch.co.za/2000/01/21/editoria/CHIEL.HTM, accessed August 19, 2009.
21. NAN, LKM R. Summers, Bulawayo, March 10, 1950, to Major Trollope.

22. NAN, LKM R. A. R. Bent, Maun, September 18, 1952, to Trollope.
23. NAN, LKM J. Desmond Clark, Livingstone, July 21, 1952, to Major Trollope.
24. NAN, LKM L. Trollope, Katima Mulilo, October 31, 1952, to Tim Brent.
25. NAN, LKM L. Trollope, Katima Mulilo, September 12, 1952, to Carp.
26. NAN, LKM D. Clark, Livingstone, November 26, 1952, to L. Trollope.
27. http://www.dispatch.co.za/2000/01/14/editoria/CHIEL.HTM,   accessed August 19, 2009.
28. NAN, LKM 1/1/3 1946–1953, Katima Criminal Cases, No. 2004/53.
29. NAN, LKM 1/1/3 1946–1953, Katima Criminal Cases, No. 2004/53. Statement made by Anderson Luendo in the presence of A. L. H. Weller, Magistrate Sesheke District, November 6, 1952.
30. NAN, LKM 1/1/3 1946–1953, Katima Criminal Cases, No. 2004/53. See Colenbrander's calls for the deportation of Luendo to Mafeking as he is worried about his and his wife's safety in Katima Mulilo.
31. NAN, LKM 1/1/3 1946–1953, Katima Criminal Cases, No. 2004/53. Statement made by Anderson Luendo in the presence of A. L. H. Weller, Magistrate Sesheke District, November 6, 1952.
32. NAN, LKM 1/1/3 1946–1953, Katima Criminal Cases, No. 2004/53. Statement made by Anderson Luendo in the presence of A. L. H. Weller, Magistrate Sesheke District, November 6, 1952.
33. NAN, LKM 1/1/3 1946–1953, Katima Criminal Cases, No. 2004/53. Statement made by L. F. W. Trollope, January 22, 1953.
34. NAN, LKM 1/1/3 1946–1953, Katima Criminal Cases, No. 2004/53. Statement made by L. F. W. Trollope, January 22, 1953.
35. NAN, LKM 1/1/3 1946–1953, Katima Criminal Cases, No. 2004/53. Statement made by L. F. W. Trollope, January 22, 1953.
36. NAN, LKM 1/1/3 1946–1953, Katima Criminal Cases, No. 2004/53. Warrant of Committal Katima Mulilo, March 10, 1953.
37. NAN, LKM Naturellekommissaris, Oostelike Caprivi Zipfel, Januarie 8, 1958, to Sekretaris van Naturellesake, Pretoria (JBG's translation).
38. NAN, LKM Naturellekommissaris, Oostelike Caprivi Zipfel, Januarie 8, 1958, to Sekretaris van Naturellesake, Pretoria.
39. NAN, LKM Magistrate Katima Mulilo, March 23, 1961, to District Commissioner Sesheke.

References

Alexander, W. J. R. (1999) "Science, History and the Kasikili Island Dispute," *South African Journal of Science* 95(8): 321–325.

Amnesty International Namibia (2003) *Justice Delayed Is Justice Denied: The Caprivi Treason Trial.* Windhoek: Amnesty International.

Boyd, J. M. (1992) *Fraser Darling in Africa: A Rhino in the Whistling Thorn.* Edinburgh: Edinburgh University Press.

Bunting, B. (1969) *The Rise of the South African Reich.* London: Penguin.

Fisch, M. (1996) *Der Caprivizipfel während der deutschen Zeit, 1890–1914*. Cologne: Rüdiger Köppe Verlag.

Green, L. G. (1961) *Great North Road*. Cape Town: Howard Timmins.

Hardy, P. & M. D. Firestone (2009) *Botswana & Namibia*. London: Lonely Planet Publications.

Hayes, P. (1996) "'Cocky' Hahn and the 'Black Venus'. The Making of a Native Commissioner in South West Africa, 1915–1946," *Gender and History* 8(3): 364–392.

Hodson, A. W. (1912) *Trekking the Great Thirst: Sport and Travel in the Kalahari Desert*. London: Fisher Unwin.

Logan, R. F. (1972) "The People of the Eastern Caprivi," *S.W.A. Jaarboek 1972*, pp. 149–152.

Macmillan, H. (2005) *An African Trading Empire: The Story of Susman Brothers and Wulfsohn, 1901–2005*. London: I.B. Tauris.

Miescher, G. (2007) "Visualising Space and Colonial Settlement: The So Called 'Farm Map' as an Icon of Settler Historiography," Paper presented at the second AEGIS conference, Leiden, July 12.

Morton, B. C. (1996) "A Social and Economic History of a Southern African Native Reserve: Ngamiland, 1890–1966," Ph.D. Thesis, Department of History, Indiana University.

Perry, A. (2000) "Caprivi Strip: World Court Awards Island to Botswana; International Court of Justice Case Concerning Kasikili/Sedudu Island (Botswana/Namibia)," *IBRU Boundary and Security Bulletin*.

Santcross, N. Ballard, S. & Baker, G. (2009) *Namibia Handbook*. London: Footprint Travel Guides.

Tlou, T. (1985) *A History of Ngamiland 1750 to 1906: The Formation of an African State*. Gaborone: Macmillan.

Trollope, L. F. W. (1956) "The Eastern Caprivi Zipfel," *The Northern Rhodesia Journal* III(2): 107–118.

Zeller, W. (2009) "Danger and Opportunity in Katima Mulilo: A Namibian Border Boomtown at Transnational Crossroads," *Journal of Southern African Studies* 35(1): 133–154.

CHAPTER 5

The "Victorian Internet" Reaches Halfway to Cairo: Cape Tanganyika Telegraphs, 1875–1926

Neil Parsons

Introduction

The telegraph wire—that forerunner of civilization.

(Hensman 1900: 135)

And what do you think of the triumph, or almost triumph, of the Cape-to-Cairo scheme from the natives' outlook?

(Question posed by anti-imperialists to Sir Harry Johnston, quoted in Weinthal 1922: 87–88)

In 1998 Tom Standage published *The Victorian Internet*, which hailed the early telegraph as the forerunner of the Internet, creating a nineteenth-century network of instantaneous electronic communication used by bureaucracies and businesses across the world. McNeill & McNeill (2003: 5–8) and van Dijk (2006: 23) have similarly argued that the Global Web of modern communications can be dated back to the invention of the telegraph 160 years ago.[1]

The Internet and cell-phone revolution of the past two decades has been greeted as a liberatory moment in universal history. Similar sentiments were being expressed a century and a half ago. In 1858, when opening the first submarine cable under the Atlantic, the US President sent the following telegram message to the British monarch, though it took 16 hours to transmit his 99 words (Headrick & Griset 2001: 549):

> May the Atlantic telegraph, under the blessing of heaven, prove to be a bond of perpetual peace and friendship between the kindred nations, and an instrument destined by Divine Providence to diffuse religion, liberty, and law throughout the world.
>
> (Kennedy 1971: 729)

Submarine cables also caught the imagination of the great imperial poet Rudyard Kipling in a collection of poems dated 1889–1896:

The Deep-Sea Cables

The wrecks dissolve above us; their dust drops down from afar
Down to the dark, to the utter dark, where the blind white sea-snakes are.
There is no sound, no echo of sound, in the deserts of the deep,
On the great gray level plains of ooze where the shell-burred cables creep.
Here in the womb of the world—here on the tie-ribs of earth
Words, and the words of men, flicker and flutter and beat
Warning, sorrow and gain, salutation and mirth
For a Power troubles the Still that has neither voice nor feet.
They have wakened the timeless Things; they have killed their father Time;
Joining hands in the gloom, a league from the last of the sun.
Hush! Men talk to-day o'er the waste of the ultimate slime,
And a new Word runs between: whispering, "Let us be one!"

According to an article in *The Economist* (2009: 139–140), the early tele-graph gave rise to new forms of "simpler, starker" thought and expression: "a new writing style, to the point and neutral in tone (or what is now called 'telegraphic')." This influenced public speaking: "short sound bites became popular because they were easier for stenographers to transcribe, and cheaper and quicker for reporters to transmit." In the words of another journal, in 1918, contrasting fast-moving films with slow stage plays: "Modernism calls for abbreviated action."[2]

The early telegraph was vital to worldwide capitalist expansion through the electronic exchange of cable news and commercial intelligence. As a French official said in 1900: "England owes her influence in the world perhaps more to her cable communications than to her navy. She con-trols the news, and makes it serve her policy and commerce in a mar-vellous manner." The imperial historian Paul Kennedy (1971: 738, 748) adds that this resulted in world telecommunications becoming "centred on

London . . . [and] increased that city's predominance in banking, insurance, business and press agency matters; it was an advantage not lightly to be thrown away." The early history of undersea exchange telegraphs features the British supercapitalist, John Pender, a former Manchester cotton merchant whose Eastern Telegraph and Anglo-American Telegraph conglomerate between the 1860s and 1911 owned half the submarine cables in the world and laid many such cables for owners of other nationalities (Headrick & Griset 2001: 559–563).

Telegraphs merely brushed the mercantile coastal fringes of Africa until the 1870s-1880s explosion of capital and labor-intensive industry (mining) into the interior of Southern Africa. Telegraphs rapidly penetrated from the coast up wagon roads, rivers, early railways and even elephant tracks. The Scramble for Africa in the 1880s-1890s was the first great "shock of modernity" for the interior of the continent. Telegraphs and railways enabled the imposition of colonial administration and coordinated the military suppression of dissent.

This chapter concerns the history of colonial telegraphy in the interior of Southern and South Central Africa up to the First World War. The telegraph was the main facilitator of efficient administration for the "subimperialism" of the British South Africa Company, which was a private corporation "chartered" by the imperial government in London to colonize new territories north of Cape Colony. Its dynamic head was the Kimberley diamond millionaire and Cape Colony politician Cecil Rhodes, whose ambition was to spread the company's rule and influence as far north as possible from Cape Colony (hence "Rhodesia"). Trunk telegraph lines were constructed through the territories seized and owned by the BSA Company (Southern, North Western, and North Eastern Rhodesia) or within the company's commercial sphere of operations (Bechuanaland and Nyasaland), including a line projected northward as part of a grand Cape-to-Cairo project along Lake Tanganyika toward Uganda.

Together with the railway, the telegraph was the outward and visible emblem and instrument of conquest and modernity facilitating white settlement, and as such was resisted at times by local African actors. On the other hand, telegraphs and railways provided channels that enabled local political actors to exploit the contradictions between "colonial" interests in Africa and "imperial" humanitarian interests beyond Africa.

The First Cape-to-Cairo Telegraph Scheme: 1875–1879

The twenty-first century undersea broadband telecommunication cables that land at Durban via Mombasa and at Cape Town via Lagos follow precedents set 130 years earlier. The first east-coast cable reached Durban from Aden via

Zanzibar in 1879, and the first west-coast cable reached Cape Town via West Africa in 1889.

Competing with the undersea cables, there was also a grand scheme dating back to 1875 for an overland telegraph line stretching from the Cape of Good Hope in South Africa to Cairo in Egypt (Weinthal 1922: 211). The practicality of the scheme had been shown by the Australian transcontinental line stretching almost 2,000 miles out of Eastern Telegraph's submarine cable station at Darwin, which was started in October 1870 and completed in August 1872.[3] The New Zealand explorer Kerry Nicholls enthused in 1876 about the day that would come when grateful Africans would regard "the thin wire that suspended a few feet above their heads . . . as the dawn of a new era of enlightened prosperity" (Raphael 1936: 48). Nile explorer Samuel Baker, when considering the value of copper wire, was more cautious about thievery: "I do not think that any police protection would protect a wire of gold from London to Inverness" (*Ibid.*: 57).

With the Cape-to-Cairo telegraph in mind, Scottish engineer James Sivewright was appointed to manage the Cape Colony's telegraph system in 1877. But his plans for the Cape Colony to subsidize the construction of a telegraph line from Pretoria to Gondokoro in the Sudan were nipped in the bud by the rapid laying of the Aden-Zanzibar-Durban submarine cable by John Pender's Eastern Telegraph Company in 1879 (*Ibid.*: 60–62; Casada 1974: 250; Wilburn 1982).

Bechuanaland Telegraph Construction 1884–1891

In 1873 the Cape Colony government had taken over the private telegraph lines (dating from 1860 onward) that ran between Cape Town and the Eastern Cape, at the same time as the colonial government of Natal bought out the Durban-Pietermaritzburg line. Cape government telegraphs were extended northward from Cape Town to the mushrooming diamond city of Kimberley in 1876. Within the next two to three years, the telegraph systems of the British colonies of the Cape and Natal and the Boer republic of the Orange Free State were linked, and the South African telegraph network was subsequently extended to the new goldfields of Johannesburg, in the Boer Republic of the Transvaal, in the 1880s.[4]

The first extension of a telegraph line north of Kimberley into Bechuanaland and the Kalahari was done by the imperial British army. In 1884–1885 Britain sent its so-called Warren Expedition from Cape Colony northward into Bechuanaland and the Kalahari to establish a new colony and a protectorate, blocking Boer-German contact across the Kalahari between German South West Africa and the Transvaal.[5] This ultramodern expedition included a mounted troop of telegraph engineers that laid a line from Barkly West (near Kimberley) to Kanye via Vryburg and Mafeking, "as rapidly as the troops marched and indeed was often in advance." In total, 225 miles of line were laid on locally acquired wooden poles in 37 days (Royal Engineers Museum n.d.; Royal Engineers Signals n.d.). The Mafeking-Kanye telegraph line was dismantled after the withdrawal of British forces in August 1885 but the wooden poles on the 220-mile Barkly West-Mafeking line were kept and replaced by permanent (metal) poles over the next four years.[6] And in 1889 the line was declared to be the property of the British Bechuanaland colonial government at Vryburg.[7]

Telegraph lines generally preceded railways, and were realigned along the line of rail once it was built. It was the promise of funds being already available to build a Vryburg-Mafeking railway (in the colony of British Bechuanaland) and a Mafeking-Shoshong telegraph (in the Bechuanaland Protectorate) that clinched a British royal charter for Cecil Rhodes's British South Africa Company in October 1889 (Olivier 1892: 5–7; Raphael 1936: 146–147). Monopoly commercial and administrative rights were granted for 25 years over the land that was to be named "Rhodesia" (later Zimbabwe) and, more ambiguously, over the Bechuanaland Protectorate (later Botswana). Those rights were extended north of the Zambezi in 1891.

However, two major chiefs or kings in the southern Bechuanaland Protectorate, Chiefs Bathoen at Kanye and Sechele at Molepolole, gave the rights to construct telegraphs and railways to rival concessionaires to the BSA Company. Sidney Shippard, Britain's senior administrator in the two Bechuanalands (colony and protectorate) and incidentally one of Rhodes's old Oxford friends, therefore turned to rival chiefs seeking to assert their independence from Bathoen and Sebele. He was looking for passage rights for the BSA Company, and telegraph construction labor from Chief Ikaneng at Ramotswa with territory on the Transvaal border (Olivier 1892: 15; Sillery 1965: 120–121 & 139), and assured Ikaneng:

> There is no witchcraft or magic in the Telegraph, there are only small holes made in the veldt to carry the poles which are about the size of a small tree, and a wire is carried along from pole to pole. There are 20 poles to a mile. That [there is] a man is at one end of the wire, and there is a box with chemicals in it, the man taps, which causes a vibration of the wire, which appears at the other end, that so many taps mean a letter and so the message is transmitted.[8]

On behalf of the BSA Company, the colonial government supervised the building of a railway (which was completed in December 1890) from Kimberley to Vryburg, Shippard's base in the colony of British Bechuanaland,

and began telegraph construction out of Mafeking toward Ramotswa on May 19, 1890 (Bolze 1968: 47), supervised by Messrs. Standford and J. A. Smith who were seconded from the Cape Colony Department of Posts and Telegraphs. Telegraph construction was speeded up when the railway reached Taung just south of Vryburg, carrying 80 tons of telegraph material.[9] Iron telegraph poles and copper wire from Siemens in England were carried onward by ox-wagon to Zeerust and then across the Ngotwane River at Ramotswa, before continuing north through Mochudi (Morkel 1985: 39).[10] The Mafeking to Ramotswa telegraph office (77 miles) was completed on June 16, 1890, Ramotswa to Palla Camp through Gaborone and Mochudi (114 miles) on August 16, and Palla telegraph office to Palapye (98 miles) on October 14.

Ikaneng's men were engaged to build the line as far as Palla Road, where it entered the vast territory ruled by Chief Khama of the Bangwato people, whose new capital was Palapye. But the line running close to the Transvaal border had to cross the territory of Chief Linchwe, based at Mochudi, whose people wanted to assert their independence from Chief Sebele as well as from the British and the Transvaal Boers. Chief Linchwe's brother Segale refused Standford labor and permission to build a telegraph office; and at "the very spot that Mr. Standford had already chosen as the site for an office," a case of telegraphic materials on a supply wagon was broken open.[11] Linchwe himself was subsequently apologetic: regretting that without a telegraph office he would remain ignorant while rival chiefs to the north and south would "hear all the news" from the telegraph.[12] Britain's high commissioner in Cape Town entertained the idea of ceding troublesome Linchwe's territory to the Transvaal Republic, until it was realized that this would cut into the line of telegraph and its accompanying main road.[13]

Fearing that there would be war with Linchwe's people, Shippard's hothead police chief Fred Carrington placed a new police camp on Linchwe's southern border. Fort Gaberones (Gaborone) was pegged out on August 24, 1890 at a crossroads called Mashoweng, next to the village of a tributary under Chief Sechele called Gaborone,[14] Old Sechele had no objection[15] but his son and successor Sebele continued to protest at the location of the new fort and at the fact that the telegraph line ran through the lands of Chiefs Gaborone and Linchwe whom he regarded as tributaries: "I said that the Telegraph line could pass along the border . . . so that it should be evidence to define the line between the Boers and myself . . . and I now find it passes through my country."[16]

Meanwhile, the Kimberley-Vryburg telegraph was being realigned along the new railway, using the existing iron poles belonging to the local colonial government to carry "two wires for Railway purposes and the ordinary Bechuanaland wire." The BSA Company maintained a base telegraph office at Mafeking that was responsible for the line through the Bechuanaland Protectorate as well as in Rhodesia. The Bechuanaland government's postmaster-general at Vryburg did, however, recognize "the danger of friction, which is always liable to arise when a system of dual control is maintained" and pleaded for the line inside the Protectorate to be removed from the BSA Company and handed to his department.[17]

The 1889 sweetheart deal between the British South Africa Company and the British government was beginning to drain the Bechuanaland colonial government's finances. In July 1890 the Vryburg postmaster-general complained about the unfairness of the BSA Company sending a mass of official telegrams free of charge along the Kimberley-Mafeking government-owned line, while the Bechuanaland government paid £1,000 a year to the BSA Company (until 1914) in return for the privilege of sending a dribble of official telegrams north of Mafeking—out of a total annual revenue of less that £2,000. But his objection was overruled by the Colonial Office in London[18] although it was agreed that the costs of a post and telegraph officer at Palapye in the Bechuanaland Protectorate should be shared equally.[19]

The BSA Company telegraph northward from Palapye was opened as far as Fort Macloutsie on May 13, 1891 and to Fort Tuli on May 28, where it passed into Rhodesia, reaching the Nuanetsi River crossing (97 miles) on September 9 and Fort Victoria on October 31.[20] The first telegram sent when the telegraph line reached Fort Tuli was from Major Leonard to Chief Khama at Palapye at 3:45 p.m. on May 28, 1891, seeking intelligence on Matabele troop movements before the telegraph line could be pushed on into southern Matabeleland (Leonard 1896: 231). The telegraph also proved of some use for the imperial British administration in the Bechuanaland Protectorate. Direct telegraphic communication was established with Chiefs Ikaneng at Ramotswa and Khama at Palapye—to the detriment of Chiefs Bathoen at Kanye and Sebele at Molepolole who lived some 30 miles off the telegraph line on the old main wagon route and whose representations to government were thus preempted.

In 1891 telegraph construction saw the first recorded industrial labor strike in the region. Workers recruited among Khama's people at Palla Road and Palapye had continued to construct the telegraph line through Matebeleland, under Mr. Smith of Cape Colony Posts & Telegraphs, but had withdrawn their labor before the line reached its terminus at Fort Salisbury. On September 26, 1891, Major Leonard met a "party" of them at the Tokwe River,

who had been working on telegraph construction, on their way back to Pilapiwe [sic]. It seems they have refused to go any further, so Smith informed me, as they must return to till their crops. Neither promises nor threats have been of any avail, and, even at the risk of incurring Khama's anger and displeasure, they have insisted on going back.

To Smith, who moved heaven and earth to dissuade them, they replied, "It is all very well for Khama and you white men. You have people who will till your lands, and take care of your crops and families in your absence, but we have no one; and if we do not do it ourselves, no one will do it for us, and our wives and children will starve and die." An answer containing much soundness and common sense, unanswerable in fact, to which Smith could not reply; and so they have gone, and left the construction of the telegraph in a serious predicament, for the intention now is to carry it right up to Salisbury.

(Leonard 1896: 311)[21]

Telegraph Construction into Rhodesia 1891–1894

The initial legal basis for "Rhodesia"—the new name that Cecil Rhodes was anxious should stick to "that part of the country" (June 1891)[22]—consisted of a concession given by King Lobengula of Matabeleland (western Zimbabwe) over the goldfields of Mashonaland (eastern Zimbabwe). The new colony was supplied by a new wagon road and accompanying telegraph wire that skirted the south of Matabeleland from Fort Tuli on Khama's frontier via Fort Victoria to Forts Charter and Salisbury in Mashonaland. The BSA Company telegraph line reached Fort Salisbury (61 miles) on February 16, 1892, thus completing a total distance from Mafeking of 772 miles.[23] The line had cost £70,000. Telegrams from London to Salisbury were charged at a bit less than half a pound sterling per word (Thy 2002: 3). In the words of one historian, "From most points of view the telegraph line . . . was more important than the fast conveyance of mails" (Keppel-Jones 1987: 348). It was essential to BSA Company commercial as well as military operations in Rhodesia. Rural telegraph stations became the bases for trading stores (Gisbourne 1967: 39) selling the consumer goods that spread the monetized economy essential to a capitalist labor market.

Fort Macloutsie, in Khama's country on the telegraph line between Palapye and Fort Tuli, became the main colonial post office in the Bechuanaland Protectorate. It served Bulawayo, Lobengula's capital in Matabeleland, with mail carried by postal runners via the previously remote mining center of Tati (near later Francistown). Macloutsie's postmaster and self-described "telegraphist," J. E. Symons, took up his post in March 1892. Symons reported on other telegraph clerks up and down the line to Mafeking and Salisbury: "There are some 'duffers' on the line, and some very smart clerks" (Symons 1892–1994: 26).

The Cape Colony's Postmaster-General took over management of British Bechuanaland posts and telegraphs on April 1, 1893, and oversight of BSA Company telegraphs in the Bechuanaland Protectorate and Rhodesia on July 1. The telegraph system reportedly consisted of "12 offices, 862 miles, [and] 60 chains of line of single wire."[24] During the Anglo-Ndebele War (actually a Rhodesian-Ndebele War) of 1893–1894, Tati was connected to Macloutsie by a government-owned light telegraph line from Palapye "of wooden poles procured locally," which was completed in January 1894.[25] In October that year the railway from Vryburg to Mafeking, built by the BSA Company, was completed and the existing telegraph line was rerouted along the railway and doubled to accommodate a railway wire (Bolze 1968: 47).

> For convenience sake the accounting was undertaken at Mafeking, where the terminal office of the Company's line was located, the Company's officials occupying a separate room in the building used as a Post and telegraph office by the Bechuanaland Government.[26]

But despite being under both Cape Government management and oversight, this dual system of ownership of the BSA Company line north of Mafeking and of the British Bechuanaland line south of Mafeking posed increasing problems concerning coordination from 1894 onward.

Rhodes Adopts the Cape-to-Cairo Telegraph

The Cape-to-Cairo idea was revived on the British imperial agenda in 1888 by colonial administrator Sir Harry H. Johnston, who was pushing British interests around Lakes Nyasa (Malawi) and Tanganyika (Raphael 1936: 409). Failing to impress the imperial government, Johnston planted the idea in the mind of Cecil Rhodes, who became prime minister of the Caper Colony in 1890 (Oliver 1957: 153). Rhodes, however, remained indifferent to the idea until 1892 when Britain's Liberal government under Gladstone wanted to retreat from the colonization of Uganda. Rhodes was fired by the idea of extending British South African or Rhodesian subimperialism and commercial power into East Africa. In November that year, he persuaded the London shareholders of the BSA Company to "build a telegraph to Uganda" (Hole 1926: 383) (Figure 5.1). To this end, the African Transcontinental Telegraph Company (ATTC) was floated as a subsidiary of the BSA Company.[27]

Under the supervision of Harry Johnston, "by December 1893 the track of the line had been cut all the way from Blantyre to Tete, and poles and wires were on their way up the Zambezi" (Oliver 1957: 243–244). But Rhodes

Figure 5.1 The Colossus of Rhodes. British press response to Cecil Rhodes's announcement of his Cape-to-Cairo transcontinental telegraph scheme. Cartoon by Edward Linley Sambourne in *Punch*, or the London Charivaria, December 10, 1892.

quarreled with Johnston: "I am not going to create with my funds an independent King Johnston over the Zambezi" (Oliver 1957: 236) and then with Sivewright, who felt slighted and ignored as "the original conceiver of the idea" (Weinthal 1922: 217). Johnston resigned in 1896 after the telegraph arrived in Blantyre from Salisbury, as he significantly saw it as bringing local colonial administration under too much direct control from London (Oliver 1957: 266), while Sivewright became a leading opposition politician against Rhodes in the Cape parliament.

African Resistance to Rhodesian Telegraphs 1892–1896

The BSA Company was happy to exploit modern technology in order to, as it saw it, bamboozle ignorant natives. In 1891, some of Lobengula's *indunas* (councilors) in Bulawayo suspected "some dark and ulterior design" behind the extension of the telegraph from the south:

> They one and all blurted out their suspicions by declaring that we whitemen were bringing up the wire to tie up their king with. They shook their heads when the "real and actual purpose" of the telegraph was explained, accepting it as truthful but declaring: "Still, it is a bad sign, this wire—a bad sign!"
>
> (Leonard 1896: 120)

Lobengula appears to have prohibited his people from stealing copper wire from the telegraph by announcing "when the telegraph line was being erected that no [person] would be able to pass underneath it without being killed." But the writ of King Lobengula was ineffective in neighboring Mashonaland, where Shona people not only were disappointed "to find that the white man's magic had no effect upon their enemies" (Kane 1954: 84) but also saw the copper wire as too tempting not to be refashioned into ornaments (Tomes 1998: 20).

The pretext for the BSA Company declaring the Anglo-Ndebele War of 1893–1894 was the theft of copper wire by Shona people on the Nuanetsi River on the Mashonaland-Matabeleland frontier. When Lobengula sent soldiers to punish the Shona in question, the BSA Company treated this as an invasion of their territory. The Matabele were subjected to a two-pronged attack by BSA Company forces from Mashonaland and imperial forces from Bechuanaland. The former column arrived in Bulawayo first and eclipsed the latter from official Rhodesian history (O'Mahoney & O'Mahoney 1963: 28).

Following the conquest of Matabeleland, the BSA Company extended a new telegraph line from Tati that reached Bulawayo (108 miles) on July 9, 1894. Further construction from Bulawayo in the direction of Salisbury then commenced (Hole 1926: 336). Hence in the words of an 1895 publication: "No sooner was this effected than receipts at the rate of 3000*l* a year were taken at the Buluwayo post office" (Knight 1895: 131). A line was also strung from Bulawayo to Fort Charter by the end of the year, thus putting Salisbury in direct contact with Bulawayo, and a start was made on replacing wooden poles with iron ones on the original line from Macloutsie via Fort Victoria.[28] In 1895, the Palapye-Tati line was improved with iron poles; better wooden poles were erected beyond Tati; and the Bulawayo-Salisbury line was also improved with better earthing and insulators. Telecommunications between Rhodesia and the south now had two alternative routes: through Tuli and Tati.[29] What had begun as a single trunk line had begun to sprout into a network.

Telegraphs played an essential role in the BSA Company's suppression of the Matabeleland and Mashonaland "risings" of 1896–1897. Most importantly, Bulawayo's telegraph lifeline to Kimberley and Cape Town via Tati was never cut. Why not? The contemporary colonial explanation was that "the Matabele [Ndebele] were terribly afraid of the telegraph wire, and hesitated to go near it, much more attempt to cut it" (Hensman 1900: 85). Closer analysis shows that the Bulawayo-Salisbury line was indeed cut by Ndebele at the Shangani River east of Bulawayo,[30] But Kalanga people along the Tati road and telegraph line southwest of Bulawayo remained impartial in what they saw as a fight between rival new and old masters (Ranger 1967: 183–190). Subsequent news of the war's end, which was carried by the telegraph, enabled some Cape Town shareholders of the BSA Company to make a "killing" in share dealings before the news reached the London Stock Exchange (Keppel Jones 1987: 501).

Serious damage was done to the telegraph of the ATTC in the Mashonaland risings against BSA Company rule and white settler occupation. A line construction supervisor was killed north of Mazoe (Burke 1971:6) and an attack on Mazoe mine resulted in the deaths of two other telegraph employees (Pollett 1957: 30).[31]

Turning the Tables on BSA Company Telegraphs 1895–1896

The telegraph was, of course, a two-way form of instantaneous communication that could on occasion do to as much harm as good for its owners. It was a carrier of news and correspondence, but also constituted a record of such dealings that could be used as evidence. This was the case in June 1895 when Chief Khama learned from the telegraph—as well as from cabled reports from London carried in Cape newspapers—about Rhodes's hurried plans to incorporate the whole of Bechuanaland Protectorate into Rhodesia. (To the south, British Bechuanaland was to be incorporated into Cape Colony at the same time.) Within days of a new Conservative government replacing the Liberal one in Britain, Khama, with missionary help, organized his brother chiefs, Sebele and Bathoen, to travel with him to London to petition against Rhodes for the continuation of the British protectorate.

After gaining widespread political support on a campaigning tour of Britain, the three Bechuana (Tswana) chiefs fatally delayed Rhodes's plans for the Jameson Raid—to use the Bechuanaland Protectorate as the "jumping off point" for a surprise attack on the Boer Republic of the Transvaal. Rhodes cabled his London agents on November 12, 1895: "It is humiliating to be utterly beaten by these niggers." Ten days later, he was still smarting when he cabled: "I do object to being beaten by three canting natives."[32] The full story is told in my book *King Khama, Emperor Joe, and the Great White Queen: Victorian Britain through African Eyes* (Parsons 1998), which shows how apparent "collaboration" with imperialism could be more effective for resistance to colonialism than overt acts of rebellion.

The Jameson Raid is a story replete with telegraphic connections. Rhodes apparently tried to stop the Jameson raiders at the last moment, by telegram. But his company secretary in Cape Town "is said to have falsified and delayed the transmission of telegrams from Rhodes to Jameson forbidding the Raid . . . (as he) had been busily feathering his own nest on the London Stock Exchange, where he was acting as a 'bear' on a large scale" (Johnson 1940: 238–239). The raiders also failed to cut the telegraph across the border connecting Zeerust to Transvaal government headquarters in Pretoria because, it is said, the two men assigned got drunk on the way and cut some farm fencing instead (Hensman 1900: 63). The youngest Jameson raider was a foul-mouthed 16-year-old telegraphist "released by the Boers on account of his youth" (Smart 1962: 22–23). A telegram sent by Germany's Kaiser Wilhelm II to the Transvaal's President Paul Kruger congratulating the latter on having repelled the raiders caused anti-German popular outrage in Britain (Hensman 1900: 64). Winston Churchill and others have seen this as the start of the deteriorating British-German relations that led up to the First World War.

Among the effects captured by Transvaal forces from the Jameson raiders, inside a black metal box belonging to Captain Robert White, was a copy of MacNeil's *General Telegraphic and Mining Code* in which a sheet of paper was found containing the cipher used in all secret telegrams. *Die Trommel van Bobbie White* (Bobby White's trunk) was to feature prominently in legal procedures as evidence of the conspiracies behind the Jameson Raid (Jameson Raid Enquiry n.d.: 251 & 256).[33] Colonial Minister Joseph Chamberlain responded immediately to news of the raid's failure by rushing to the Eastern Telegraph's offices in London and seizing all copies of BSA Company telegrams. At the subsequent parliamentary enquiry, as the reading out of telegrams progressed, "Dr. Jameson could not conceal his distress and mortification . . . he flushed deeply, even his ears turning a blazing scarlet" (Jameson Raid Enquiry n.d.: 256–259). Radical MP Henry Labouchere suggested that some "missing telegrams" would have implicated Chamberlain himself in the Jameson Raid conspiracy.

Telegraph Administration in the Rhodesias and Bechuanaland 1897–1910

The telegraph north of Mafeking was rerouted along the new railway and pressed ahead at full speed during 1896–1897 because of the African "risings" in Rhodesia and the near impossibility of ox-drawn wagon traffic during the Rinderpest livestock epidemic. The railway via Gaborone reached Mochudi on March 1, 1897, Palapye Road on July 1, Francistown (Monarch Mine on the road to Tati) on September 1, and its terminus at Bulawayo on November 4, 1897 (Croxton 1982). The old telegraph line from Palapye to Macloutsie and Tuli was kept and connected by an extra 11 miles to the main trunk line at Palapye Road railway station.[34] The telegraph line along the railway was increased to three wires—one was "reserved for railway purposes, two others 'for public use'."[35]

The burden of telegraphic traffic in 1896–1897 proved too great for the Mafeking exchange. The BSA Company temporarily transferred its exchange from Mafeking to the company's headquarters at Kimberley. After Fort Macloutsie failed to replace Fort Gaborone as the main police base in the Bechuanaland Protectorate in 1894, Macloutsie was progressively abandoned as a fort.[36] But the telegraphists at Macloutsie, Palapye Road, Palla, and Gaborone were recruited by the BSA Company (later the Southern Rhodesian civil service), which maintained the telegraph lines in an uneasy working relationship with Cape Colony posts and telegraphs.[37]

BSA Company posts and telegraphs became independent of Cape Colony posts and telegraphs administration from February 23, 1897, with the appointment of its own postmaster-general. It took over direction and maintenance of its own telegraph system to as far south as Mafeking.[38] But the BSA Company refused to join the South African Telegraphic Union, which would have meant charging usage rates as low as those of the Transvaal and Cape Colony, and instead placed surcharges on incoming telegrams. The Cape postmaster-general objected strongly to the continuation of the BSA Company privilege of free telegraph passage between Mafeking and its Kimberley offices at Cape government expense. Unsatisfactory relations between the two systems had been exacerbated by the dramatic increase in Rhodesian traffic in 1896–1897:

> Although the Cape Colony performs the bulk of the work, the Company receives by far the largest proportion of the revenue earned... As a matter of fact the Cape Colony performs... twenty operations... as compared with

eight operations by the Company, notwithstanding the latter receives four times the amount of revenue accruing to the Colony.[39]

In 1901 Southern Rhodesia and the Bechuanaland Protectorate joined the Universal Postal Union, and thus the South African Telegraphic Union. As a result, from January 1, 1902, the BSA Company reduced its telegram tariffs from South Africa to two pence a word (minimum two shillings) to the Bechuanaland Protectorate and Southern Rhodesia, three pence to Beira railway offices, and five pence to African Trans-Continental offices.[40]

Rhodesia's first telephone exchange opened in Salisbury in 1898 and was staffed exclusively by white female telephone clerks (Kane 1954: 229), in contrast with the employment of mission-educated black male clerks in colonial Africa to the north. According to the 1911 edition of the *Encyclopaedia Britannica*, the telegraph and telephone system of Southern Rhodesia was "very complete . . . There being for the whole of [Southern] Rhodesia about 8,000 m[iles] of wires. The total includes the police telephone wires and part of the Transcontinental system, and is served by about ninety telegraph offices" (*Encyclopaedia Britannica* 1911). However, an effective Salisbury-Bulawayo telephone trunk line sturdy enough to carry the traffic of numerous private business and domestic branches was only established in 1914 (Kane 1954: 229).

Within the Bechuanaland Protectorate a new telegraph line from Palapye Road to Serowe was built at the request of the white traders in the Serowe Chamber of Commerce in 1903–1904. The BSA Company wanted to limit the line to a telephone service, to avoid the expense of employing a trained telegraphist at Serowe. But traders objected that commercial secrets would not be blabbed over the telephone. The Company was eventually persuaded to pay half the salary of the Serowe postmaster to act also as a telegraphist. In the south of the Bechuanaland Protectorate, the Union of South Africa's post office was persuaded to lay a branch line from Zeerust in the Transvaal to Lobatse in Bechuanaland in the 1920s, which was afterward extended to Kanye.

The BSA Company and its post-1923 successor, the government of Southern Rhodesia, continued to depend on the multichannel telegraph along the railway through the Bechuanaland Protectorate as Rhodesia's main wire to the Cape. Fraught overlapping connections between Rhodesian and South African telecommunication authorities were to persist until 1962 when the Bechuanaland Protectorate government finally took control of telecommunications within the country.

In April 1905 the Rhodesian railway and telegraph were extended from Southern into North Western Rhodesia (western Zambia) with the completion of the Victoria Falls Bridge by the Cleveland Bridge Company from

County Durham (UK). The railway and telegraph reached the mining center of Broken Hill (Kabwe) in July 1906 and were extended to Bwana Mkubwa mine (Ndola) in November 1909.[41] By July 1910, the railway reached Sakania in Katanga where it linked up with a short line constructed by mining companies from Elizabethville (Lubumbashi) in the Belgian Congo (Hunt 1959: 13). Telegraph and mail services in North Western Rhodesia were under the control of a comptroller of posts and telegraphs stationed first at Kalomo and then at Livingstone, the capital of the colony from 1911 until 1935 when Lusaka became the capital of Northern Rhodesia (Ridley 1955: 16).

Nyasaland and the African transcontinental company telegraph 1896–1926

Despite the sabotage of the ATTC line in Southern Rhodesia in 1896, construction continued along Lake Malawi and the road that marked the (later) Zambia-Tanzania border. Local labor and food supplies were hard to come by and the company fought short wars before local people were "pacified," that is, induced to render labor tribute. But in 1899, the ATTC telegraph reached the south end of Lake Tanganyika and a one-mile-square lakeshore base was set up at Kasakalwe (Gamwell & Gamwell 1961: 520).

At the beginning of 1899, Cecil Rhodes personally concluded an agreement with the Kaiser in Berlin whereby Germany would allow the telegraph line up the eastern side of Lake Tanganyika to British Uganda (Hensman 1900: 128–131; Gann 1965: 156). But the Cape-to-Cairo project effectively died with the death of Rhodes on March 26, 1902. The line of telegraph was surveyed as far as Kisumu, the terminus of the Uganda Railway on Lake Victoria, but construction actually reached no further than Ujiji, halfway along the eastern shore of Lake Tanganyika. It was completed in 1903 thanks to African carrier laborers supplied by the Germans (Denny 1962: 41).[42]

The ATTC proved to be a liability with limited utility. Its poles were made of vulnerable wood and maintenance was made difficult by crocodile-infested swamps, dense woodland and ravines (Denny 1962: 41). Along Lake Malawi, "apart from the natives (who coveted the wire for bangles and the like) elephants were the worst trouble makers" (Twynam 1953: 54). The ATTC line was effectively little more than a third telegraph network for the colonial government of the underdeveloped Nyasaland Protectorate (Malawi),[43] complementing the government's own telegraph system and the railway-telegraph link to the Eastern Telegraph submarine cable out of Quelimane in Mozambique.

The death knell for the ATTC was sounded in 1911 by the opening up of Marconi wireless telegraphy across the Sudd in the Sudan, which helped decide Chartered Company investors to abandon it (Denny 1962: 42). It was decided in the same year to liquidate the company, which had accumulated losses borne by the BSA Company of £95,900 by 1910. The ATTC line in Northern Rhodesia and German East Africa was ripped apart by enemy action during the First World War when a temporary military telegraph line was strung between trees by British military forces from Broken Hill up what was to become known as the Great North Road. But the line inside Nyasaland was "maintained and kept open by the Liquidator" (Weinthal 1922: 217) until 1926 when the Nyasaland government posts and telegraphs department acquired full possession. The African Transcontinental Telegraph then passed out of living memory.[44]

The Rise of Wireless Telegraphy (Radio)

The first Marconi "wireless telegraphy" stations were opened at Durban and Cape Town in 1910. By 1915, Marconi had three high-powered (30 kilowatt) wireless stations on the South African coast, connecting with Mombasa and Cairo and Leafield in Oxfordshire, England (Weinthal 1922: 437–441). But the Marconi network was purely for communicating with ships at sea. There was limited use of such radio on both sides during the 1914–1918 campaign in German East Africa. The British contented themselves with short-range field wireless for coordination between units (Langham 1959: 258). The Germans tried to relay upper atmospheric radio messages from Berlin to East Africa. One very significant radio message in December 1917 stopped the Zeppelin airship L 59, just south of Khartoum, from flying on with ammunition and medical supplies for the German army in East Africa (Langham 1959: 264–265; Cree 1968: 84).

The Marconi company's effective monopoly of wireless communication within the British Empire and at sea was not broken until after 1919. It was replaced by government post-office monopolies over wireless signal transmission and by state broadcasting corporations. Elsewhere in North America and Europe, private radio transmission oligopolies flourished. By 1928, because of the fear that "the cheaper wireless services might force the more strategically valuable cable companies out of business," the British government promoted the amalgamation of British undersea cable companies into Cable & Wireless Ltd. The latter was eventually nationalized in 1950 to maintain the intercontinental security of Britain's submarine communications (Kennedy 1971: 750).

Cheaper wireless communication enabled an intensification of colonial administration in Africa from the later 1920s onward, at the same time as colonial administrative procedures were being standardized (Furse

1962: 57–58, 62–86, 134–162). Coordination between the European colonial powers was reflected in the 1935–1938 founding of the African Telecommunications Union.[45] In the case of the Bechuanaland Protectorate, an official radio or wireless network exchanging abbreviated administrative correspondence ("Savingrams") between district radio stations, manned by the police, was established in 1936. The main transmitter at Mafeking also became a public broadcaster in the early evening, transmitting government news and music to loudspeakers that were placed in the central courtyards of major villages (Zaffiro 1991: 2).

Conclusion

Telegraph cables and overhead wires have survived from Victorian times, even if telegrams were generally superseded by telephones and radio. They have recently been restored to international importance by the Internet and the World Wide Web. We have seen what is, in effect, a reversion from wireless telegraphy (radio) to cable telegraphy, as was predicted in part by a lecture given in 1918:

> Speaking at a meeting of the Stoll Picture Theatre Club in London, Mr. Low Warren, one of the pioneers of the film industry, predicted that before long cinematography would be harnessed up to other great scientific inventions such as the telegraph, the cable, and wireless. It would even be within the bounds of possibility to cable a short film many thousands of miles, and we might see pictorial representations of great events in New York or Johannesburg. It might even be possible in the near future by means of wireless rays to show a topical picture simultaneously in a thousand theatres at the moment of occurrence.[46]

This chapter has shown how telegraphs were transmitters of early colonial power and nowhere more so than in the huge tract of Southern Africa where, for a decade, a single commercial enterprise, the British South Africa Company, held sway. In a mere seven years, the indigenous people of Zimbabwe were conquered by violence and stripped of their land, leaving a grievance of unusual vehemence that has persisted (and has been politically exploited) to the present day.

Telegraphic communications were able, however, to check administrative abuses as well as summon the forces of repression. The historian Paul Kennedy (1971: 751) has pointed out that the Aden-Durban cable was rapidly laid in 1879 for "amongst other reasons to supervise [High Commissioner] Sir Bartle Frere's activities at the Cape." Telegraphs and regular mail instituted long-distance bureaucratic control from the Colonial Office in London over the colonial district official who was previously "administering his district from his lodge in the wilderness and from his tent . . . not tied to an office chair and edicts and demands from the capital or from London" ("R. H. Palmer" 1959: 187).[47] Old "Bobo" Young, who endured the transition in Northern Rhodesia (Zambia), continued to send back official questionnaires, sent by colonial headquarters, blank with a terse note saying: "Leave it to the man on the spot!"[48] Not that some administrators did not learn to exploit the system. R. V. Kubicek has observed how telegraphic language "gave local officials the chance to exaggerate any crisis" (Kennedy 1971: 751).

Perhaps the most widespread and insidious impact of the telegraph was as the timekeeper of industrial capitalism. Colonial society was now regulated everywhere by the telegraphic transmission of the temporal tyranny of Cape Mean Time, one and a half hours ahead of Greenwich Mean Time (South African time has since been adjusted to two hours ahead), dictating not only hours of employment and trade but even leisure, down to the duration of a football match (Keppel-Jones 1987: 579).

With the extension and transformation of telegraph trunk lines, carrying direct orders, into telephone networks promoting dialogue between parties, colonial society itself became more complex. Within Southern Africa, the intensity of early telegraph and telephone networks was a direct corollary of the extent of white colonial settlement, urban and rural. The strategic nodes of this network were the telephone exchanges. It is scarcely surprising that the white worker revolutionaries of the 1922 Rand Rising in South Africa "were endeavouring to reach the Telephone exchange" as their main target.[49]

We have seen how, in 1895, some African rulers made use of the new telegraphy to appeal directly to the imperial metropole over the heads of colonial interests. But telegraph lines were considered irrelevant by the general populace until "the thickening of the global human web" made possible a "mass society marked by mass communication networks" (van Dijk 2006: 23). The sheer expense per word of sending telegrams limited their utility, but public usage went up as rates came down. Similarly, telegraph lines were converted to carry telephone calls. The pioneer possessors of telephones among Africans were chiefs and members of chiefly families, and members of the emergent urban elite:

> The fixed line telephone was a status symbol . . . In homes that had telephones, the moment the instrument rang, it was not promptly answered. It would be allowed to ring for a long time in order to make sure that everybody knew that particular home was blessed with a telephone.

People who lived in town but did not own a telephone were under immense pressure. They had to justify to their country bumpkin cousins why they did not own a telephone.

(*Loose Cannon* 2009)

Widespread use of telephones disconnected, adapted, or perpetuated old connectivities between people and land, the rulers and the ruled, and between families and individuals, and new linkages and connectivities came to dominate the region.

Notes

1. Outside Europe and North America, there has been scholarly writing on the history of early telegraphs in India (Karbelashvili 1991), in China (Knuesel 2007) and Australia (Blainey 1966/1988). Late-colonial amateur historical studies of early telegraphs in Malawi and Zambia (Twynam 1953; Denny 1962; Smith 1975; Baker 1976) do not appear to have been matched south of the Zambezi.
2. *Stage and Cinema* (Johannesburg), April 28, 1917, p. 4.
3. Cape Colony Postmaster-General's Report for 1898: 64–65 reproduced in Thy 2001: 30.
4. www.telecom.co.za/aboutus/history/index.html. See also Houghton & Dagut (1972: 122).
5. "The object of this mission and expedition is to remove the filibusters from Bechuanaland, to pacificate the territory, to reinstate the Natives in their lands, to take such measures as may be necessary to prevent further depredations, and, finally, to hold the country until its further destination is known" (memorandum by Warren dated October 29, 1884 & quoted in BPP—C.4227 of November 1884 & C.4588 of August 1885 despatch no. 47, p. 13).
6. Thy (2001: 7), citing Cape Colony Postmaster-General's Reports for 1885: 22 & 1886: 12.
7. Thy (2001: 8) citing Cape Colony Postmaster-General's Report for 1889: 19.
8. Report of a meeting between Sir Sidney Shippard and Chief Ikaneng, Ramotswna, May 12, 1890, as recorded by G. D. Smith of the Kanya Concession Company—copies in Botswana National Archives, PRO—CO 417/47 despatch no. 834 (Secheleland Concession) & Confidential Print African No. 441.
9. *Ibid.* despatch no. 632 ("Telegraph construction in the Protectorate") Sivewright to High Commissioner's Office, August 18, 1890.
10. Thy (2001: 9) citing Cape Colony Postmaster-General's Report for 1890: 16–17; Thy (2002: 8); Symons (1892–1894: 16).
11. PRO—CO 417/45 despatch no. 658 ("Interference of Bakhatla with telegraph extension") Fuller to Carrington, August 4, 1890; Thy (2001: 9 & 11 citing Cape Colony Postmaster-General's Report for 1890: 16–17).
12. *Ibid.* despatch no. 658 Fuller to Carrington, August 4, 1890.

13. PRO—CO 417/46 despatch no. 762 ("Cession of a portion of Protce to S.A.R.") minute by Graham.

14. See also PRO—CO 417/45 despatch 659 ("Site for fort near Gaberone's town") Administrator Vryburg telegram to High Commissioner Cape Town, August 25, 1890. The historic original telegram is in Botswana National Archives File HC 1470.

15. PRO—CO 417/46 confidential despatch (CO 20033 rcd. October 13, 1890): Surmon to Shippard, September 3, 1890.

16. *Ibid.* despatch no. 861: "Sechele's objection to telegraph & well sinking."

17. PRO—CO 417/47 despatch no. 868 (Annual Report for year ending September 30 [1890]'), Report of the Postmaster General and Superintendent of Telegraphs.

18. RO—CO 417/46 despatch no. 725: PMG Vryburg to Shippard, July 11, 1890; Shippard to Loch, July 18, 1890, Harris to Imperial Secretary Cape Town, September 11, 1890, & Loch to Knutsford September 13, 1890. See also Cape Colony Postmaster-General's Report for 1896: 74–79 reproduced in Thy (2001: 25–26).

19. *Ibid.* despatch no. 776 ("Post & Telegraph office at Palachwe") Postmaster-General Middleton Vryburg to Secretary Asburnham Vryburg, September 3, 1890.

20. Thy 2001: 11–12 citing Cape Colony Postmaster-General's Reports for 1890: 16–17 & 1891: 27.

21. Most accounts underline the self-confidence of the Bangwato in this period to the annoyance of putative white colonists. The first crime recorded in the *Pioneer Corps Order Book*, on May 15, 1890, was that of a Mongwato named Muicemong (Moetsemong) for inciting another Mongwato to refuse duty. He was fined 15 shillings (Johnson 1940: 130).

22. Sir Henry Loch to Lord Knutsford, June 16, 1891, quoted in *Rhodesiana*, vol. 38 (March 1978), p. 77. The name Rhodesia seems to have originated in Cape Colony newspapers in the latter part of 1890 (Gray 1956: 77).

23. Thy (2001: 11 citing Cape Colony Postmaster-General's Report for 1891: 38).

24. Thy 2001: 13–15 citing Cape Colony Postmaster-General's Report for 1893: 16 & 39.

25. After the manager of the Tati mine, William Frank Kirby, complained bitterly about the insecurity of the mining settlement in the face of the oncoming Anglo-Ndebele war (BPP—C.7171 of September 1893 despatch no. 65, Imperial Secretary Cape Town to Secretary Tati Concession Company, Kimberley, July 29, 1893; despatch no. 67, p. 70, Manager Tati mine to Assistant Commissioner Palapye, August 3, 1893); Thy (2001: 15 citing Cape Colony Postmaster-General's Report for 1893: 39).

26. Cape Colony Postmaster-General's Report for 1897: 74–79 reproduced in Thy (2001: 24–26).

27. Hensman (1900: 124); Rollin (1913: 1978); Raphael (1936: 152–153); Denny (1962); and Weinthal (1922: 215–216).

28. Thy 2001: 15–16 citing Cape Colony Postmaster-General's Report for 1894: 20 & 30–32.

29. Thy 2001: 17–18 citing Cape Colony Postmaster-General's Report for 1895: 38–39 & 44–45.

30. Cape Colony Postmaster-General's Report for 1896: 49–50 reproduced in Thy (2001: 21).

31. For other accounts of the Mazoe telegraphists, see Howland (1963: 20) which suggests the message to help was Major Forbes up the ATTC line; Hodder-Williams (1967: 53 n. 46) noting that the telegraph was not cut until nearly a week after Mangwende's people killed the African missionary Bernard Mizeki; and Wood (1985: 85) with Mrs. Cass's running commentary from her viewpoint in the laager on the two men being killed outside apparently quoted verbatim. Blakiston was referred to as "an unsuccessful kind of intellectual" (Keppel-Jones 1987: 480).

32. Parsons (1998: 222–223) quoting W. T. Stead, *The scandal of the South African committee: A plain narrative for plain men* London: Review of Reviews Office, 1899: telegrams 14, 18 & 26.

33. Bobby White's action was thus explained, if not exonerated, by newspaper responses reviewing Hugh Marshall Hole's *Jameson Raid* published in 1926. (White's signed copy of that book, containing newspaper clippings and with a comment thanking that he still had friends, was found by me in an Edinburgh antiquarian bookstore and has since been donated to the Oppenheimer Brenthurst Library in Johannesburg.)

34. Cape Colony Postmaster-General's Report for 1897: 23–24 reproduced in Thy (2001: 23; 2002: 7).

35. Thy (2001: 19–21 citing Cape Colony Postmaster-General's Report for 1896: 20, 37–38, 45–50).

36. Cape Colony Postmaster-General's Report for 1897: 74–79 reproduced in Thy (2001: 25–26); PRO—CO 417/126 confidential despatch ("Police Headquarters") Gool-Adams to Shippard, August 23, 1894; Administrator Vryburg to Governor, September 4, 1894; Acting Governor Cameron to Ripon, September 10, 1894.

37. BPP—C.7962 February 1896 (*South Africa: Correspondence Relative to the Visit to this Country of the Chiefs Khama, Sebele, and Bathoen, and the Future of the Bechuanaland Protectorate*) despatch no. 49, p. 45, Officer Commanding Macloutsie to Resident Commissioner Vryburg, November 26, 1896; p. 49, Acting Secretary J. A. Stevens British South Africa Company, Cape Town, to Imperial Secretary Cape Town, December 2, 1895.

38. Cape Colony Postmaster-General's Report for 1897: 74 reproduced in Thy 2001: 24.

39. Cape Colony Postmaster-General's Report for 1897: 74–79 reproduced in Thy (2001: 27–28); Cape Colony Postmaster-General's Report for 1898: 79 reproduced in Thy (2001: 29).

40. Cape Colony Postmaster-General's Report for 1902: 30–31 reproduced in Thy (2001: 35).
41. Jordan (1957: 201) confirms that the telegraph from the south was in use at Bwana Mkubwa/Ndola in 1910.
42. See photo in Weinthal (1922: 216).
43. In 1900 there were ten telegraph stations in Nyasaland (Chikwawa, Chiromo, Blantyre, Domira Bay, Kotakota, Bandawe, Nkhata Bay, Isisya, Florence Bay, Karonga), one in Southern Rhodesia (Inyanga), four in Northern Rhodesia and one in Tanganyika (Bismarksburg) (Denny 1962: 41).
44. See the ignorance of the otherwise well-informed Kenneth Bradley (1959: 2) in an article originally published in 1938, claiming that "to this day, Abercorn remains the terminal section."
45. See *African Telecommunication Agreement* (1935).
46. "Film by cable," *Stage and Cinema* vol. vii, no. 165, 28 September 1918, p. 5.
47. *A Lodge in the Wilderness* was John Buchan's influential 1906 novel in which barely disguised personalities debated the colonial administration at a symposium in Kenya.
48. *Northern Rhodesia Journal*, vol. 2, no. 2, 1954, p. 48.
49. *South African Pictorial, Stage and Cinema* (Johannesburg), vol. xiv, no. 347, March 25, 1922, p. 19.

References

African Telecommunication Agreement, with Annex, Telegraph Regulations, and Final Protocol of the Telegraph Regulations (1935) Geneva: League of Nations (Société des Nations—Recueil des Traites, 1938, no. 437, pp. 53–82). Accessed November 10, 2009: http://untreaty.un.org/unts/60001_120000/19/20/00036953.pdf.

Baker, C. (1976) The administration of posts and telecommunications 1891–1974. *Society of Malawi Journal* 29(2): 6–33.

Blainey, G. (1966/1988), *The tyranny of distance: How distance shaped Australia's history*. Melbourne: Macmillan.

Bolze, L. W. (1968) The railway comes to Bulawayo. *Rhodesiana* 18: 47–84.

Bradley, K. (1959) Adventurers still. *Northern Rhodesia Journal* 4(1): 1–8, part 4 of an article in *African Observer* (1936).

Burke, E. E. (1971) Mazoe and the Mashona rebellion, 1896/97'. *Rhodesiana* 25: 1–34.

Casada, J. A. (1974) James A. Grant and the Royal Geographical Society. *Geographical Journal* 140(2): 245–253.

Cree, A. (1968) Memories of the 1914–18 campaign. *Northern Rhodesia Journal* 3(5): 465–466.

Croxton, A. H. (1982) *Railways of Zimbabwe: The story of the Beira, Mashonaland and Rhodesia Railways*. 2nd edition. Newton Abbot: David & Charles.

Denny, S. R. (1962) The Cape to Cairo telegraph. *Northern Rhodesia Journal* 5(1): 39.

Economist (2009) 393(8662): 138–140: Network effects. How a communications technology developed.

Encyclopaedia Britannica, 1911 edition, Entry on "Rhodesia." Accessed October 14, 2009: www.1911encyclopedia.org.

Furse, R. D. (1962) *Aucuparius: Recollections of a recruiting officer*. London: Oxford University Press.

Gamwell, H. & M. Gamwell (1961) The history of Abercorn. *Northern Rhodesia Journal* 4(5): 515–524.

Gann, L. H. (1965) *A history of Southern Rhodesia: Early days to 1934*. London: Chatto & Windus.

Gisbourne, G. G. (1967) Memoirs of D. G. Gisbourne: Part 1 Australia to Salisbury: 1890–1892. *Rhodesiana* 17: 34–55.

Gray, J. A. (1956) A country in search of a name. *Northern Rhodesia Journal* 3(1): 75–77.

Headrick, D. R. & P. Griset (2001) Submarine telegraph cables: Business and politics, 1838–1939. *Business History Review* 7: 543–578.

Hensman, H. (1900) *A history of Rhodesia: Compiled from official sources*. Edinburgh & London: William Blackwood.

Hodder-Williams, R. (1967) Marandellas and the Mashona rebellion. *Rhodesiana* 16: 27–54.

Hole, H. M. (1926) *The making of Rhodesia*. London: Macmillan. Reprinted Bulawayo: Books of Rhodesia (1967).

Houghton, D. H. & J. Dagut (comps) (1972) *Source material on the South African economy 1860–1970: Volume 1, 1860–1899*. Cape Town: Oxford University Press.

Howland, R. C. (1963) The Mazoe patrol. *Rhodesiana* 8: 16–33.

Hunt, B. L. (1959) Kalomo-Livingstone in 1907. *Northern Rhodesia Journal* 4(1): 9–17.

Jameson Raid Enquiry (n.d.) "Cipher messages" being chap. xxii, pp. 248–269, from unidentified source on of House of Commons Committee, reproduced in *The Victorian Dictionary* compiled by L. Jackson. Accessed October 27, 2009: www.victorianlondon.org/publications2/lookeron-22.htm.

Johnson, F. (1940) *Great days: The autobiography of an empire pioneer*. London: G. Bell & Sons. (Facsimile reprint by Bulawayo: Books of Rhodesia, 1972).

Jordan, W. K. (1957) Memories of abandoned bomas no. 11: Old Ndola'. *Northern Rhodesia Journal* 3(3): 200–202.

Kane, N. S. (1954) *The world's view: The story of Southern Rhodesia*. London: Cassel.

Karbelashvili, A. (1991) Europe-India telegraph "bridge" via the Caucasus. *Indian Journal of History of Science* 26(3): 277–281.

Kennedy, P. M. (1971) Imperial cable communications and strategy, 1870–1914. *English Historical Review* 86(341): 728–752.

Keppel-Jones, A. (1987) *Rhodes and Rhodesia: The white conquest of Zimbabwe, 1884–1902*. Kingston & Montreal/Pietermaritzburg: McGill-Queen's University Press/University of Natal Press.

Knight, E. F. (1895) *Rhodesia of today: A description of the present condition and prospects of Matabeleland and Mashonaland.* London: Longmans Green.

Knuesel, A. (2007) British diplomacy and the telegraph in nineteenth-century China. *Diplomacy and Statecraft* 18: 517–537.

Langham, R. W. M. (1959) Memories of the 1914–18 campaign, part IV. *Northern Rhodesia Journal* 4(2): 166–179.

Leonard, A. G. (1896) *How we made Rhodesia.* London: Kegan Paul, Trench, Trübner & Co. (facsimile reprint Bulawayo: Books of Rhodesia).

Loose Cannon [newspaper column] (2009) The Telephone. *Sunday Standard* (Gaborone), July 5.

McNeill, J. R. & W. H. McNeill (2003) *Human web: A bird's eye view of world history.* New York: W. W. Norton.

Morkel, A. R. (1985) Early days in Mashonaland. *Heritage of Zimbabwe* [formerly *Rhodesiana*] 5: 31–59.

Oliver, R. A. (1957) *Sir Harry Johnston & the scramble for Africa.* London: Chatto & Windus.

Olivier, S. (1892) *Memorandum on the origin and operations of the British South Africa Company,* by. S. O. Downing Street, October 13, 1892. Printed as PRO—CO 879/37 Colonial Office Confidential Print, African (South) No. 439.

O'Mahoney, B. M. E. & K. E. O'Mahoney (1963) The southern column's fight at Singuesi, November 2, 1893. *Rhodesiana* 9: 28–36.

Parsons, N. (1998) *King Khama, Emperor Joe, and the great white queen: Victorian Britain through African eyes.* Chicago: University of Chicago Press.

Pollett, H. (1957) The Mazoe patrol. *Rhodesiana* 2: 29–38.

Ranger, T. O. (1967) *Revolt in Southern Rhodesia: A study in African resistance.* I London: Heinemann.

Raphael, L. A. C. (1936) *The Cape to Cairo dream: A study in British imperialism.* New York: Columbia University Press.

Smith, J. (1959), R. H. Palmer. *Northern Rhodesia Journal* 4(2): 187–188.

Ridley, H. C. N. (1955) Early history of road transport in Northern Rhodesia. *Northern Rhodesia Journal* 2(5): 16–23.

Rollin, H. (1978) *Rollin's Rhodesia.* Translated by D. Kirkwood from French, original 1913. Bulawayo: Books of Rhodesia.

Sillery, A. (1965) *Founding a protectorate: History of Bechuanaland 1885–1895.* The Hague: Mouton & Co.

Smart, H. W. (1962) Early days in Bulawayo (1896–1900). *Rhodesiana* 7: 22–33.

Smith, R. C. (1975) The Africa trans-continental telegraph line. *Rhodesiana* 33: 1–18.

Symons, J. E. (1892–1894), Contributions to *St Martin's-le-Grand, The Post Office Magazine* (London). Reprinted in P/Thy (ed.) (2002).

Thy, P. (2001) *The Northern Mails and Telegraphs: Bechuanaland and Rhodesia in the Annual Reports of the Postmaster-General, Cape of Good Hope* Davis, CA: Krone Publications.

Thy, P., ed., (2002) *The Macloutsie post office and its postmaster Bechuanaland Protectorate 1892*. 2nd Edition. Davis, CA: Krone Publications.

Tomes, I. (1998) The Matabele war, 1893. *Heritage of Zimbabwe* 17: 18–73.

Twynam, C. D. (1953) The telegraph in British Central Africa. *Nyasaland Journal* 6(2): 52–55.

van Dijk, J. A. G. M. (2006) *The Network Society*. 2nd Edition. London: SAGE Publications (1st edition *de netwerkmaatschappij* Houten: Bohn Staflen van Loghum, 1991).

Weinthal, L., ed., (1922) *The story of the Cape to Cairo railway and river route, 1887–1922*. London: Pioneer Publishing/African World.

Wilburn, K. (1982) "James Sivewright" *Dictionary of South African Biography*. Cape Town: Tafelberg, vol. 4, pp. 572–574. Accessed October 27, 2009: http://www.members.shaw.ca/beyondnootka/biographies/sivewright.htm

Wood, R. H. (1985) A visit to Mazowe. *Heritage of Zimbabwe* (formerly *Rhodesiana)* 5: 83–90.

Zaffiro, J. J. (1991) *From police network to station of the nation: A political history of broadcasting in Botswana, 1927–1991*. Gaborone: Botswana Society.

Archival Sources

African Studies Centre Library, Leiden British Parliamentary Papers (Command Papers, or "Blue Books") from Isaac Schapera Collection:

C.4588 *Transvaal . . . and Adjacent Territories* (August 1885)

C.5363 *Bechuanaland . . . and Adjacent Territories* (April 1888) & C.5524 (August 1888)

C.7171 *South Africa . . . Mashonaland and Matabeleland* (September 1893)

C.7962 *South Africa . . . . Bechuanaland Protectorate* (February 1896)

Public Record Office, London (National Archives of England & Wales) Colonial Office Dispatches from High Commissioner at Cape Town

CO 417/45, 46, & 47 (dispatched August-September 1890; September–October 1890; October–November 1890)

CO 417/126 (dispatched September 1894)

Colonial Office Confidential Print (African)

CO 879/34: African No. 404: *Memorandum as to the Jurisdiction and Administrative Powers of a European State holding Protectorates in Africa*

CO 879/35–37: African (South) *Further Correspondence respecting the Affairs of Bechuanaland and Adjacent Territories* No. 414: *(February 1892),)* No. 426: *(October 1892).* No. 441: *(May 1893).*

CO 879/37: African (West) No. 436: *Further Correspondence relating to Territories on the River Gambia (January 1894)*

CO 879/37: African (South) No. 439: *Memorandum on the Origin and Operations of the British South Africa Chartered Company* [by Sydney Olivier]

CHAPTER 6

Marriages and Mobility in Akan Societies: Disconnections and Connections over Time and Space

Astrid Bochow

Introduction

At the Central Market in Kumasi in Ghana I befriended two women, Auntie Emi and Sister Afia, both in their mid-thirties and living apart from their husbands who were in the United States. Auntie's husband had left her soon after they got married and the birth of their now ten-year-old son. In the first two years, he sent money but then stopped doing so and she had heard nothing for five years. The two women expressed their longing for a new husband and a second child but were faced with the situation that, as married women, they could not go out in the evening for a beer without becoming the target of gossip.

A few months after this conversation, Auntie Emi met a man and fell in love. He promised to marry her even though he was married to a woman who was living in Israel. He wanted Auntie Emi to become his second wife. After initially hesitating, the situation changed Auntie Emi's perspective on life and she decided to divorce her first husband. This was a complicated process that involved convincing first her mother and then her maternal uncle, who she used to live with when she was a child. Being Catholic, her mother denounced the possibility of divorce, reasoning that she had left her husband and that she did not want her daughter to do the same. Her family disagreed with the divorce at first and begged her to contact her husband before considering any further steps. Through her brothers and sisters who were living abroad, she was able to contact him but got no answer. At this point I left for Germany.

In Auntie Emi's experience, her relationship with her husband was a disconnected one. While her husband was abroad, she was left behind in Ghana with their only child but, unlike many other women in similar situations, she was not facing economic hardship. After finishing secondary school, she had followed in her mother's footsteps, at first helping her around her shop and eventually taking it over. She had six siblings living in Europe and the United States who would regularly send money to Ghana so her suffering was not economic but personal. Being married without a physically present husband, she felt deprived of the pleasures of womanhood that most women in the Akan setting enjoy, namely, having (more) children and enjoying a sexual partnership with a man. Having met somebody who was willing to fulfill her wishes and wanted to marry her, she had to face her family's opposition as well as that of her in-laws, who did not accept her wish to divorce.

There are many women in Kumasi like Auntie Emi who live apart from their husbands who have migrated to a country in the North to work and earn money. This situation challenges the very nature of the understanding of marriage as has been described in the ethnographic literature on Akan-speaking societies in present-day Ghana. The situation makes it necessary for married spouses to find a way to maintain their (marital) bond over time and space. In addition, Auntie Emi's story highlights the personal desire for intimate relationships, and how these have been betrayed and shaped by the situation she finds herself in.

This chapter explores how bonds are created in transnational marriages. What connections do couples who are living in opposite corners of the world use? Who can mobilize such connections and what happens when they fail? How does this situation of living separately on two different continents change the institution of marriage itself? The following ethnographical explorations offer a critical reading of the ethnographic literature on Akan marriage, which commonly stresses that marriage exists predominantly as a material structure of reciprocity. In contrast, the chapter highlights the emotions and individual aspirations in marriages. Looking into connections and disconnections in transnational marriages allows an understanding of the changes in this institution in a multilayered process to which new infrastructures, changing (Christian) moralities of the family, kinship obligations, and personal desires and aspirations contribute.

Transnational Marriages through the Prism of the Local: Methods and Context

There is a growing literature on the Ghanaian diaspora (van Dijk 2004; Krause 2008) that explores the sociality of migrant communities in host societies. Other authors have followed migration networks by pursuing a mul-tisited approach to migration (Mazzucato 2005; Nieswand 2005). In contrast to these studies, this chapter develops a perspective on transnational marriages through the prism of the local and the subjectivities of those wives who stayed behind in Ghana. The data presented here are the result of 12 months of fieldwork in Kumasi and Endwa, a rural town near Kumasi. My core research question revolved around premarital sexuality and courtship in Pentecostal churches (Assemblies of God and Lighthouse Chapel International) and secondary schools. When I was not doing interviews or attending church activities, I used to sit with the local teachers in Endwa or with market women in Kumasi. And I sometimes accompanied them when they were preparing for funerals.[1] The data I discuss here are therefore the product of long-term observation over the whole period of my research between 2004 and 2006, and the perspective on transnational marriage is gendered and highly contextual and reflects individual experiences.

Two contexts are particularly relevant for the purpose of the case studies: the context of educated families of farming background in Endwa and those from a trading milieu in Kumasi. In Endwa I stayed with a farmer, Kodjo Kuma, his wife, and two of his daughters. Kodjo Kuma used to work as a mechanic in Accra but had started working on his own farm in the mid-1980s when the company he was working for closed down. Attached to his household were his second daughter, Teresa, and her two children who lived in her own house but shared the evening meal with the rest of the family. Teresa was a teacher in the village. She rarely left the house except to go to work and lived a family-centered life. The family can be classified as having a farming background but with a strong orientation toward education and city life. In Kumasi, I stayed in the home of a well-off trading woman, Auntie Paulina. She was the oldest of seven siblings and had built a two-story house with her brothers and sisters. She lived there with her mother, husband, children and their partners and children, as well as various siblings and nephews and nieces who helped her in her business and who she supported financially. Auntie Paulina, like Auntie Emi, had taken over her business from her mother and other maternal relatives. The market women organized their lives in a female-oriented sociality: they moved in a group of female friends and relatives who they spent their time with and they helped each other to prepare funerals and other important occasions. These market women were quite wealthy by local standards and exhibited a strong orientation toward matrilineal acquisition and the distribution of wealth.

In both contexts, the women exhibited a strong dedication to Christian values and even though they did not all belong to a Charismatic movement, they listened to radio programs featuring famous (Charismatic) preachers.

They embraced the message of the "prosperity gospel" they heard and stressed the importance of leading a holy and decent life. In both contexts I met married women whose husbands, as well as other relatives and friends, were living abroad.

A Society on the Move: Globalization and Migration in Ghana

Ghanaian society has undergone tremendous changes in the past two decades. Since the 1990s, the liberalization of the media, the country's relative economic growth, an increase in transatlantic migration, and a rise in the number of Pentecostal churches have had a huge impact on Ghana's social life. Global influences have come to penetrate Ghana's public and private life and, hand in hand with its increasing wealth, infrastructure has also improved. For example, there are now Internet cafes in the urban centers, more rural areas have electricity, and the Ghanaian mobile-phone provider, Spacenet, has extended its network to cover the rural areas. Interpersonal communications as well as trade have been enhanced by these improvements in infrastructure.

Historically, the kingdom of Asante and the ethnic groups on the coast were well-connected by trade and migration with other ethnic groups nearby, as well as with Europe at the beginning of the seventeenth century. In the colonial period, labor migration from neighboring African countries to the cocoa fields and the mines in present-day southern and central Ghana were part of life for the people in this region (Adepoju 2005: 30–32). From the mid-1970s onward, the country experienced years of economic hardship and import restrictions under General Acheampong, which led to a severe shortage of goods (Oquaye 1980: 17). On top of all this, there was a drought in 1982/1983. The recession, which was caused by the liberalization of the economy, affected living conditions in Ghana in the 1980s and 1990s (Buah 1998: 219) and economic hardship and political pressure, as a result of military rule, triggered a wave of emigration to the so-called countries of the North (Higazi 2005; Martin 2007: 56–57). Ghana Immigration Service figures on outmigration show that between 1990 and 2000 the number of people migrating from Ghana to Europe increased twofold, and from Ghana to the United States it rose tenfold.[2] The number of Ghanaians living abroad has been estimated at 21 million, with large Ghanaian communities in the diaspora having been founded in global centers such as London, Hamburg, New York, Toronto, and Washington (Peil 1995; Anarfi et al. 2003: 3; Bump 2006). Of importance are not only people's strong emotional bonds and religious ties to their home regions but also the remittances they send that contribute significantly to Ghana's economy (Ter Haar 1998; van Dijk 2002; Tiemoko 2004; Nieswand 2005). The Central Bank of Ghana has estimated that remittances amounting to US$1.2 billion account for the country's second most important source of foreign currency and cover the education and healthcare costs of many families (Bump 2006).

Kumasi, and especially the milieu of trading women, is renowned for hosting a "culture of migration." I met several people at the Central Market there, both men and women, who had been living in the Netherlands, Germany, or the United Kingdom. In a random sample of 15 market women, 9 had close relatives, such as a brother or husband, living abroad. Going abroad to earn money and coming back to "be a somebody" is common in Ghanaian society and has become a type of lifestyle among young men and, to a lesser extent, among young women too (Martin 2007). A local TV show entitled "Greetings from Abroad," which was broadcast on GTV between 2004 and 2006, featured Ghanaian migrants in various countries demonstrating not only the global spread of the Ghanaian diaspora but also how much migration is a part of everyday life in Ghana. It is not unreasonable to say that Ghanaian society is a "society on the move."

A good example of how household arrangements and family survival are structured around migration was the household in Kumasi where I was staying. The head of the household, Auntie Paulina, was a successful trading woman. She was the oldest of her seven siblings and although her sisters helped her in the business, the young men had migrated to the United Kingdom to study, work, and earn money to start businesses for themselves. One of the young men in the household said: "It makes me sad to think about my former class mates. They have all gone abroad." This feeling of having been "left behind"—as well as left out—and being alone in Ghana was shared by many young people in their twenties and early thirties.

Accommodating Akan Marriages: Mobility and Residence Patterns

The ethnographic literature on marriages in Akan societies concurs that the institution of marriage is of little social function in the setting of matrilineal kinship arrangements (cf. Clark 1984: 103). In the first half of the twentieth century, Akan marriages were described as consisting of duo local arrangements (Rattray 1929; Fortes 1950/1975; Bleek 1976; Allman & Tashjian 2000). According to Akan matrilinear understanding of kinship, an adult woman and her children would rather belong to her mother's brother's (*wofa*) household than to her husband's. She would prefer to work on her brother's farm than on her husband's, and a husband and wife would live in separate households and meet only in the evenings when the wife came to her husband's house to prepare food. Given this situation, the bonds of marriage would not be created through living together and running a common household but would be maintained through sharing food and a bed. Marriage is thus constituted through a certain immediacy of interaction and exchange. The realization of this union depends very much on day-to-day interaction in which flows of goods and services as well as the flow of bodily fluids constitutes and consolidates the relationship. This is reflected in the concept of marital love that revolves around "care" (in Twi: *Hwe no yie*). Care has a gender-specific division of labor according to which the husband looks after his wife and children, and the wife cooks food for her husband, washes his clothes, and cleans his house (see Oppong 1980; Clark 1999).

This understanding of marriage is reflected in the rules according to which spouses can divorce. A man has the right to divorce his wife if she does not cook for him or sleep with him; and a wife can divorce her husband if he does not provide her and her children with food and/or if he beats her. Ethnographic observers from the 1930s to the 1980s noted the marital bond in Akan societies as being easy to dissolve if these criteria are no longer met, and have described a divorce as easy to obtain, since only the council of family elders has to be called upon. The claim has to be spelled out and the elders then dissolve the marriage (Fortes 1950/1975; Clark 1984: 106).[3] In addition, researchers have stressed the fluid nature of Akan marriages, which are formed in a process rather than marked by one point of entry, such as the wedding ceremony (Fortes 1950/1975; Allman & Tashjian 2000: 75, 77). According to customary law, a marriage needs to be conducted in three consecutive ritual steps with the first, the knocking (*kokoko*) declaring the "intention" to marry and the last marking the final and binding moment of being married. In everyday practice, a knocking often follows an unplanned pregnancy and the marriage is never finalized as Akan people put it. I observed several cases in which men, even though they did not marry their partner according to custom, nevertheless, took on the responsibility of being a caring father and supporting their wives and children.

The residential patterns of married people have changed. Since the 1920s, as a result of the professionalization of cocoa farming and labor migration, women have become more attached to their husbands than to their matrilineal kin (McCaskie 2000). While moving with their husbands and working closely with them in their fields, women tended to reside with their husbands more often than before (Allman & Tashjian 2000: 9ff). Not only did the position of women change during the first part of the twentieth century but the marital role of men also altered. In the second half of the twentieth century they became their family's breadwinner and were supposed to support their wives and families (Allman & Tashjian 2000: 122–123; Miescher 2005: 124). A similar evolution was observed in other African societies at the beginning of the twentieth century (Lindsay 2003) when the opening up of the public service to African personnel as well as labor migration supported such family models (Lindsay 2003; Lindsay & Miescher 2003).

In present-day Kumasi and Endwa, household arrangements are highly flexible and facilitate the flow of people from the city to the village and vice versa. A highly mobile group of educated farmers of "rich" crops, such as cocoa or palm fruits, have a lifestyle that allows them to spend periods of time in the city as well as in the rural areas depending on the season, the need for labor in the rural household, and their activities in the city. Many people switch between rural contexts and life in the city and move in and out of relatives' and friends' households due to education or in order to find a better job. If they lose their business in the city, they may move back to the rural area. Alternatively, they may get divorced or be separated and be looking for a new place to live, or they may simply be on bad terms with the people in the house they were living in. Sometimes whole families have several places to live: they might stay at a place near their work during the day, and sleep at another place at night. The problem of accommodation is solved in various ways and "houses" and "housing" have different meanings in different people's lives. Fortes (1950/1975) considered the wish to own one's own house an issue that concerns "every man" in Ashanti. van der Geest (2000), too, has shown that houses are symbols of prestige and success. The family house in one's mother's home region, which is usually headed by an older relative, is the place family can turn to for accommodation and food.[4] If a person dies, relatives will do everything to complete an unfinished building for the funeral. In the Christian setting, "homeliness" is attached to new attributes. According to the prerogatives of the church, the husband *in spe* has not only to prove that he can provide for a certain standard of housing but also has to have his own room, which should be furnished with a couch, chairs, and a TV. The home is also considered to be a spiritually safe and secure place (for Latin America, cf. Martin 1995: 101).

In this highly mobile social setting, there are several reasons for a husband and wife not to reside together. Sometimes appropriate accommodation is lacking and housing in urban areas is becoming an economic problem for young people as living space has to be rented and rents often need to be paid two to three years in advance.[5] For women, there are other reasons not to live with their husbands. Their parents may be aging or sick and need their care or they may have just given birth for the first time, which means that a woman usually goes to her mother's household where she can expect help with the birth itself and with taking care of the baby. For educated professionals, it is common to live separately, especially if both partners are working in the public sector and in different locations. And affordability, comfort, convenience and better working conditions may account for the decision not to live together. Living separately is the reality for many couples in Accra, Tema, Endwa, and Kumasi, and is having an impact on the emotions of those involved. It often leads to tension and conflicts between spouses, with most issues revolving around questions of trust. On both sides, the idea that the spouse is planning his/her moves separately frequently leads to distrust.

In such situations, children appear to create a bond between their parents but could also be the source of conflict. They present an effective way for women to connect with their husbands or the father of their children, as has been considered by Guyer (1994) and Haram (2004), who show that marriage and children are important to women in building their lateral networks. Due to shifting male responsibilities since the middle of the twentieth century, men have increasingly been expected to take care of their wives and their own biological children, in contrast to the matrilineal understanding in which the maternal uncle takes care of any children (Oppong 1981: 54–56; Clark 1999; Miescher 2005: 124).[6]

Building Bonds for Life: Marriage, Migration, and Family Planning

Not only do individual life plans center around migration, they are also part of the planning young men and women do together, of premarital relationships. In Christian settings in particular, a relationship may start out with a common goal, that is, to get married. Marriage requires preparation and money. Partners discuss and plan how and when they will proceed with their wedding, how much money they will need, how they will live and when they will be able to reach their goals of establishing a family, building a house, and financing their children's education (Bochow 2010: 255–259). For many, migration to the West offers the opportunity to build a lifestyle that can create a position of respect in society, namely, owning a house, a car, and a business. Migration shapes not only life courses but also histories of relationships: both partners usually agree that the husband will migrate and work abroad but some couples plan for the wife to join her husband after a few years. Others may arrange for the wife to stay in Ghana and take care of any children they may have. Migration projects can span many years, even decades, a fact that both partners are well aware of.

Teresa and Kofi were such a couple and had planned their marriage and subsequent lives along the migration timeline. Teresa's father, Kodjo Kuma, had insisted that all his daughters should go to school and had paid the school fees by selling goats at the beginning of each semester. After finishing secondary school, Teresa attended a teachers' training college where she met her future husband. Their first son was born shortly before they got married as it took them some time to prepare financially for their church wedding. Looking at the wedding pictures, which showed Teresa in a white wedding dress, she commented that her mother-in-law was opposed to the idea that her son was getting married in church. She feared that the son would now turn away from his family and fail to support them. Shortly after the wedding, the couple decided that Kofi should go abroad because they realized that their salaries as teachers would not allow them to live the life they wanted to lead. Kofi tried to get a visa for the United Kingdom as this was known to be a relatively easy process. Shortly before their daughter Marjam was born, his visa came through and he left for the United Kingdom. Once there, he would send money home to Teresa and the children that he earned from manual work and as soon as he was able to legalize his status, they agreed, he would come and visit them.

As her husband was working abroad, Teresa could live comfortably in a well-furnished house that she owned, something she and her husband would not have been able to acquire if he had stayed in Ghana. How did Kofi and Teresa manage their marriage, the distance, and its enforced separation? And how was the flow of goods and money guaranteed?

As in any other duo local marriage arrangement, Kofi sent the bulk of the money not for his wife to enjoy but for the upkeep of their children. Teresa, for instance, explained that she used the "chop money" Kofi had sent her for the children to build a house in Endwa and she had not asked for "extra money." The arrival of the mobile-telephone network in Endwa had facilitated their communication considerably. In 2004 Teresa was among the first in the village to have a mobile phone, which was given to her by one of her brothers-in-law in Kumasi.[7] Before she had a mobile phone, she had to go to Fosso every week and receive her husband's call in a telecommunications center. Sometimes, she remembered, the call would not come and she would have to wait or leave without speaking to him. Now she was spared the trip to Fosso and could receive calls whenever they came, but they had decided to talk to each other only once a week to minimize the costs.

A last connecting moment had occurred just before I came to stay with the family in Endwa when Kofi had been able to legalize his status and had come to visit his family after being away for three years. A new stereo deck, a DVD player, and flashy children toys were the visible signs of his recent stay. He had also left Teresa pregnant. With the same determination to plan their lives and their relationship that they had shown when planning his migration, they decided that Kofi would try to visit them every year. From some of her remarks, I concluded that Teresa assumed that her husband was living with a "white woman" in the United Kingdom who had enabled him to acquire a permanent visa. This would not have been exceptional: Ghanaian men and women alike are convinced that men cannot live alone and without sexual intercourse. Ghanaian women in her situation often lived with the knowledge that their husbands would be living with another woman while abroad.

In contrast to Auntie Emi's experience of being disconnected from her husband, the connections in Teresa's case between her and her husband worked well. Their plan to acquire money to enhance their standard of living and guarantee a good education for their children was well-designed and both partners showed commitment to it. They were using all kinds of technologies—mobile communication, visa regulations, Western Union (to send money), and the husband's siblings—to maintain the flow of money and goods. Teresa is an example of a woman for whom a transnational marriage arrangement has worked out. In a study on microlevel flows of remittances, Mazzucato (2004: 6) showed that 35% of all private transactions from migrants are sent to parents, and 29% are sent to siblings. These are then spent on the care of elderly parents and on school fees. Only 3% of all transactions are received by wives back in Ghana but if a wife can count on these flows, they will be of a considerably higher amount than those received by siblings and parents. In 1999 alone, an estimated US$2,413 was paid to wives while the amount sent to parents was US$ 443, and US$ 337 went to siblings. These figures reflect the situation of Auntie Emi and Teresa. Auntie Emi represents a typical case, with a husband who does not send money but siblings who do. This had enabled Auntie Emi to lead quite a comfortable lifestyle: she buys beautiful clothes and invites friends to her family house where she cares for her elderly mother. In contrast to Auntie Emi's case, Teresa's husband was sending her considerable sums of money so that she was able to build a house. Having possession had, however, created a distance between her and her family members in Endwa, who rarely came to see her or watch TV or videos with her as I would have expected them to. Migration and the flow (or not) of luxury items and money are, however, not the only outcome of the new connections between Ghana and countries in the North. These connections, created by new technologies and infrastructure, have also enabled other developments, as the continuation of Auntie Emi's case suggests.

When I returned the following year, Auntie Emi had managed to get a divorce and as the issue had already been settled, she showed little interest in explaining what had happened. The divorce was not the only change. As she had a new boyfriend, she was spending less time at the market. He had not married her, contrary to his initial promises. Her friends had also changed. Sister Efia, who had been a regular visitor in her shop, now rarely came to visit as a result of a period of tension and disagreement revolving around business relations with a wholesaler in Accra. Sister Efia reported that when Auntie Emi had begun seeking a divorce, her husband had called her and had pleaded with her to remain married. He admitted he had done wrong and promised her US$10,000 as compensation for his long silence and to cover child maintenance. But Auntie Emi rejected his offer. Sister Efia commented that she herself was not involved with any men and would wait for her husband to return. She also believed that Auntie Emi had taken advice from the wrong people. "Her friends, they have all got their husbands here in Ghana and sleep with them. But when you get old, you'll be lonely. Some do! They come back after years. And then they live here with their own wives."

Sister Efia's perception of events took an unexpected turn. First, it represents a clear contradiction of the personal dreams and hopes she had expressed only nine months earlier. Second, it contrasts with the picture of marriage given in the ethnographic literature that stresses the importance of the immediacy of exchange in marriage relations. The atmosphere of public piety, due to the penetration of Pentecostalism in all aspects of social and public life (Meyer & Moore 2006; de Witte 2008), has resulted in the growing importance of women being married in terms of social status. In addition, women seem to be putting more emphasis on being married now than in the past. This leads to some concluding reflections on the nature of the connections people are using in this time of transnational marriages as well as the effects these connections are having on marriages.

Discussion: Understanding Connections and Disconnections in Transnational Marriage Arrangements in Kumasi

I have introduced cases of two women. They are both about the same age and both their husbands are living in a country in the North. While one of them, Teresa, is benefiting from the connections that are available in Ghanaian society in the twenty-first century, the other, Auntie Emi, is continuously disconnected from her husband, economically as well as emotionally. The flow of money and information and the ability to communicate can be perceived as a certain position of agency that one woman is occupying and the other not. One woman is able to mobilize the flow of money and goods using existing infrastructure, such as telecommunication technologies as well as a kin network to communicate. The other woman is failing to do so and has been disconnected from her husband for a long time. Why is money flowing in these transnational arrangements in one case and not in the other? And how can the agency of the one and the lack of agency of the other be described in the framework of marriage in this context?

Teresa and Auntie Emi occupy slightly different class positions. Being from a farming background, Teresa and her husband placed an emphasis on education and the couple showed a strong commitment to Christianity by marrying in church. They thus followed historical tradition: education, brought by the missions, had been a means for nonroyals to climb the social ladder in the former Gold Coast in the first half of the twentieth century (Foster 1968). Deeply engaged with Christian values, this sector of the population was pursuing a model of marriage that emphasized the ties between the spouses as well as the ties to their own biological children. This is reflected in the fact that, in the 1970s, civil servants whose fathers had been educated themselves paid school fees for their own biological children but did not feel any obligation to contribute to the upbringing of their extended matrilineage (Oppong 1981). Teresa's mother-in-law's rejection of her son marrying in church reflects this historical experience with marriages having been conducted according to Christian norms. In this context, Teresa's ability to mobilize her husband to use the existing infrastructure, such as mobile phones and international money transfers, for the steady flow of money seems to follow a historical trend according to which education and Christianity have strengthened men's financial responsibilities toward their wives and children. Due to a strong orientation to these former elite lifestyles, the success story of this couple is easy to understand in the light of historical processes of social positioning within Ghanaian society throughout the twentieth century.

Auntie Emi's background is that of an urban trading family with a strong orientation toward matrilineal kinship. In her family, just as in the trading family I was staying with, wealth is distributed through one's siblings and inheritance was handled matrilineally. It is important to note that Auntie Emi's failure has not been so much of an economic nature: Even though her husband did not send the money as promised, she did not suffer economically as her six siblings working in Europe and the United States did send money. She perceived her suffering as personal as she was bound to a man who was absent and unreachable, even when she requested a divorce. The literature on marriage in the Akan context tends to stress the provisory nature of marriages in this context where men, according to matrilineal rule, have little financial obligation toward their wives and children (Clark 1984; Mikell 1997; Allmann & Tashjian 2000). This partly reflects a common understanding of marriage among people in Kumasi themselves as they see "care," also in its material aspects, as a central aspect of marital love. Nevertheless, these authors tend to focus on the material nature of marriage and see the fulfillment of maintenance as evidence for a well-functioning marriage. This literature seems to ignore moral as well as emotional and personal factors that might contribute to the success or failure of marriage. In this case success means the ability to mobilize transnational flows in marriage, and failure is the lack of such mobilization. It would, however, be too easy to highlight merely the subjective elements of desire for a fulfilling relationship, which brings out the experience of disconnection in Auntie Emi's case. This would ignore other important factors that contributed to Auntie's situation, namely, the moral power her kin (her mother, siblings, maternal uncle as well as her in-laws) exercised when they encouraged her to wait for her husband's approval before divorcing him. On the other hand, these kin networks are crucial to Auntie Emi's economic strategies and they enable communication as Auntie Emi was only able to connect with her husband through their mediation. Auntie Emi's experience of disconnection therefore shows a complex picture in which emotions and personal desires for a (sexual) partner as well as the morality of kinship and economic interdependencies contribute to her situation. The perspective of Auntie Emi's history proposed by her friend Efia offers an even more complex picture. It suggests that the experience of transnational marriages, with people leaving the country and coming back decades later, impacts on expectations of the institution of marriage. Alongside Christianity, it is contributing to a new conjugal morality in the context of migration.

Conclusion

With today's global influences, life is changing in southern Ghana and the new infrastructure is offering access to economic growth through migration. On the other hand, as migration becomes the rule and many people are experiencing separation from their spouses, siblings, and parents, people are increasingly seeing what it is like to be disconnected in a social and economic sense, but also emotionally. The question is no longer whether people are globally connected or not, as migration, media images, and Western-styled products are present in everyday life in Kumasi. It is instead how to connect to or with these influences materially and emotionally. The situation of couples living apart on two continents would seem to open up new possibilities of agency for some as it gives them access to economic resources that they can use for building houses, buying cars, and providing education. For others, it revolves in a situation of powerlessness. The two cases show that there are many factors and a multilayeredness of processes involved whereby certain individuals gain agency and are able to use existing infrastructures to create connections, while others feel powerless. Historical elements, such as Christianity, have offered the normative framework for a parent-centered family model as well as a lived elite praxis. The latter has strengthened paternal responsibilities toward children, as well as the personal capacities of women, such as their dedication to strict time plans of phone calls and visits over the years. Other elements include the interplay between kinship as a source of moral power, an economic network, and a network of communication.

Finally, marriage itself has changed in this multilayered process. The experience of living apart for years, maybe even decades, is common for many married couples. The possibility of connecting through modern telecommunication technology, money transfers, and occasional visits and the experience of disconnection in combination with the possibility of reconnecting after a long period of separation are changing the normative framework of marriage itself. This change is influenced by social moralities regarding what married women are supposed to do or prefer not to do. While new infrastructures continue to accelerate economic gain and enhance its flows, transnational marriages may become a profitable enterprise in some cases. At the same time, these connections are having an effect on personal sentiments as they offer the possibility of fostering the hope of fulfillment in marriage in a far distant but bright future. In such cases, marriage may become a new romantic dream of a better future.

Notes

1. The observation of lived realities of married couples, Pentecostal and non-Pentecostal, created the context for the aspirations of young people planning marriage.
2. In 1999 there were 33,425 migrants to Europe and 22,688 to the United States; in 2000 there were 56,558 migrants to Europe and 120,216 to the United States. In 2001 there were 26,698 migrants to the United States (a drastic decrease in numbers) and in 2002 36,812 migrants went to the United States (a slight decrease). Figures taken from Twum-Baah (2005: 58).
3. Kyei (1992) points out the moral impediments of divorce. If either a husband or wife wants a divorce, they have to confront their spouse before the council of elders and announce their wishes. They finalize their demands by bringing a drink of *Akpeteshi* (local gin) to their spouse's family.
4. In his memoirs, Kyei (2001), Fortes's research assistant, describes how he migrated with his family to the land of his extended kin where they intended to farm in the first half of the century. He describes their arrival and the warm welcome they received. This suggests that the migration experience was forming life courses even in the first part of the century.
5. This rule was developed in years of recession when the value of any money decreases so rapidly that fixed rents are unprofitable and constantly increasing rents would be unacceptable.

6. This was reflected in the family law in 1985, which attributed financial responsibilities to the genitor of the children.
7. The other two people who had mobile phones were the chief who owned a taxi company and was once the representative of the drivers' association in Assin Fosso, the district capital close to Endwa, and a young girl whose twin sister had won the Green Card lottery and was living in the United States.

References

Adepoju, A. (2005) "Patterns of Migration in West Africa," in: T. Manuh (ed.), *At home in the world? International migration and development in contemporary Ghana and West Africa*. Accra: Sub-Saharan Publishers, pp. 24–54.

Allman, J. & V. Tashjian (2000) *"I will not eat stone." A women's history of colonial Asante*. Oxford: James Currey.

Anarfi, J. K., S. Kwankye, O. M. Ababio & R. Tiemoko (2003) *Migration from and to Ghana: A background paper*. Legon, Development Research Centre on Migration, Globalisation and Poverty.

Bleek, W. (1976) "Sexual relationships and birth control in Ghana. A case study of a rural town," Ph.D. Thesis, University of Amsterdam.

Bochow, A. (2010) *Intimität und Sexualität for der Ehe. Gespräche über Ungesagtes in Kumasi und Endwa. Ghana*. Hamburg: LIT Verlag.

Buah, F. K. (1998). *A History of Ghana. Revised and Updated*. London & Basingstoke: Macmillan Education LTD.

Bump, M. (2006) "Ghana: Searching for opportunities at home and abroad," Retrieved from http://www.migrationinformation.org/Profiles/print.cfm?ID= 381, December 2007.

Clark, G. (1984) *Onions are my husband. Survival and accumulation by West African market women*. Chicago & London: Chicago University Press.

Clark, G. (1999) "Negotiating Asante family survival in Kumasi, Ghana," *Africa* 69(1): 66–85.

de Witte, M. (2008) "Spirit media. Charismatics, traditionalists, and mediation practices in Ghana," Ph.D. Thesis, University of Amsterdam.

Fortes, M. (1950/1975) "Kinship and marriage among the Ashanti," in: A. R. Radcliffe-Brown & D. Forde (eds.), *Systems of African kinship and marriage*. London: Oxford University Press, pp. 252–285.

Foster, P. (1968) *Education and social change in Ghana*. Chicago: University of Chicago Press.

Guyer, J. (1994) "Lineal identities and lateral networks: The logic of polyandrous motherhood," in: C. Bledsoe and G. Pison (eds.) *Nuptiality in Sub-Saharan Africa. Contemporary anthropological and demographic perspectives*. Oxford: Clarendon Press, pp. 231–255.

Haram, L. (2004) " 'Prostitutes' or modern women? Negotiating respectability in Northern Tanzania," in: S. Arnfred (ed.) *Re-thinking sexualities in Africa*. Uppsala: Almquist & Wiksell Tryckerei, pp. 211–232.

Higazi, A. (2005) *Ghana Country Study. A part of the report on informal remittance systems in Africa, Caribbean and Pacific countries (ACP)*. Oxford: Centre on Migration, Policy and Society (COMPAS).

Krause, K. (2008) "Transnational therapy networks among Ghanaians in London," *Journal of Ethnic and Migration Studies*, Special Issue "African European Linkages," 34(2): 235–251.

Kyei, T. E. (1992) *Marriage and divorce among the Asante. A study undertaken in the course of the Ashanti Social Survey 1945*. Cambridge: African Studies Centre.

Kyei, T. E. (2001) *Our days dwindle. Memories of my childhood days in Asante*. Portsmouth: Heineman.

Lindsay, L. A. (2003) "Money, marriage, and masculinity on the colonial Nigerian railway," in: L. A. Lindsay & S. F. Miescher (eds.), *Men and masculinities in modern Africa*. Portsmouth, NH: Heinemann, pp. 138–156.

Lindsay, L. A. & S. F. Miescher (eds.) (2003) *Men and masculinities in modern Africa*. Portsmouth, NH: Heinemann.

Martin, B. (1995) "New mutations of Protestant ethic," *Religion* 25(2): 101–117.

Martin, J. (2007) "What's new with the 'been-to'? Educational migrants, return from Europe and migrant's culture in urban Southern Ghana," in: H. P. Hahn & G. Klute (eds.) *Cultures of migration. An African perspective*. Hamburg: LIT Verlag, pp. 203–237.

Mazzucato, V. (2004) "The impact of international remittances on local standards: Evidence of households in Ghana." Presented at the Migration and Development Conference, Accra, July 18–20, 2004.

Mazzucato, V. (2005) "Ghanaian migrants' double engagement: a transnational view of development and integration politics," *Global Migration Perspectives* 48.

McCaskie, T. C. (2000) *Asante identities. History and modernity in an African village 1850–1950*. Edingburgh: Edingburgh University Press.

Meyer, B. & A. Moors (eds.) (2006) *Religion, media, and the public sphere*. Bloomington: Indiana University Press.

Miescher, S. F. (2005) *Making men in Ghana*. Bloomington: Indiana University Press.

Mikell, G. (1997) "Pleas for Domestic Relief: Akan Women and Family Courts," in: G. Mikell (ed.) *African feminism. The political survival in sub-Saharan Africa*. Philadelphia: University of Pennsylvania Press, pp. 96–126.

Nieswand, B. (2005) "Charismatic churches in the context of migration: Social status, the experience of migration and the construction of selves among Ghanaian migrants in Berlin," in: A. Adogame & C. Weissköppel (eds.), *Religion in the context of African migration*. Scheßlitz: Rosch Buch, pp. 243–267.

Oppong, C. (1980) "From love to institution: Indications of change in Akan marriage," *Journal of Family History* 5(2): 197–209.

Oppong, C. (1981) *Middle class African marriage. A family study of Ghanaian senior civil servants*. London: George Allen & Unwin.

Oquaye, M. (1980) *Politics in Ghana 1972–1979*. Accra: Tornado Publications.

Peil, M. (1995) "Ghanaians abroad." *African Affairs* (94): 345–367.

Rattray, R. S. (1929/1956) *Ashanti law and constitution*. Oxford: Claredon Press.

Ter Haar, G. (1998) *Halfway paradise. African Christians in Europe.* Cardiff: Cardiff Academic Press.

Tiemoko, R. (2004) "Migration, return and socio-economic change in West Africa: The role of family." *Population, Space and Place* (10): 155–174.

Twum-Baah, K. (2005) "Volume and characteristics of international Ghanaian migration," in: T. Manuh (ed.) *At home in the world? International migration and development in contemporary Ghana and West Africa.* Accra: Sub-Saharan Publishers, pp. 55–77.

van der Geest, S. (2000) "Funerals for the living: Conversations with elderly people in Kwahu, Ghana," *African Studies Review* 43(3): 103–129.

van Dijk, R. A. (2002) "Religion, reciprocity and reconstructing family responsibility in Ghanaian Pentecostal Diaspora," in: D. Bryceson & U. Vuorela (eds.), *The transnational family. New European frontiers and global networks.* Oxford: Berg, pp. 173–196.

van Dijk, R. (2004) "Negotiating marriage: Questions of morality and legitimacy in Ghanaian Pentecostal Diaspora," *Journal of Religion in Africa* 34(4): 438–468.

CHAPTER 7

A Ritual Connection: Urban Youth Marrying in the Village in Botswana

Rijk van Dijk

Introduction

Much of the current work on marriage in Africa investigates how relationships are becoming transnationalized (crossing nation-state borders) as well as transculturalized (crossing cultures) and how the commercialization and commoditization of weddings is occurring in the process. Objects are playing an ever-increasing role in wedding arrangements and seem to link local weddings with global worlds of style. Masquelier (2004), for example, demonstrates how in marriages in Niger young girls not only are attracted by the prospect of marrying a partner who has traveled but also expect particular Western commodities to be part of the marriage arrangements and the bride price. In a number of other African situations, the combination of a variety of desires in the context of arranging marriages—or "marriage-scapes" (Constable 2009)—demonstrates remarkable similarity (Johnson-Hanks 2007; Cole & Thomas 2009; Pauli 2009). In some cases, religion is becoming a mediating factor in arranging this commoditized styling of marriage that is capable of connecting different life worlds. In Ghana, for example, transnational Pentecostalism mediates marriage relations between partners and their families (van Dijk 2004; Soothill 2007; Bochow 2008, 2010). This extends into the Ghanaian diaspora overseas and such marriages are attractive, requiring Western consumer and luxury items to achieve a specific (aspired) status. While the marriage links diverse life worlds, the objects being exchanged in the process appear to have the agency to make such interconnections possible. How can the agency of objects be understood? They appear to be crucial as linking devices for the way marriage connects different life worlds, economies, and cultural domains, however far apart they may be.

Weddings all over Africa are becoming increasingly costly (Mann 1985; Cornwall 2002; Masquelier 2005; Johnson-Hanks 2007; Pauli 2009) as a result of inflation, and this is generating competition that large sectors of the population are finding hard to escape. Consumerism increases the functioning of marriage as a mark of status in a social landscape where being middle class appears as a role model by being able to connect to global styles involving luxury items that are usually only available in urban centers. The middle-class style is now being communicated from the urban to rural places via weddings. This contribution aims to demonstrate that rural places, such as Molepolole, can be a major driver in the manifestation of global styles through weddings, which traditionally have to be held in the villages.

While the term "marriage-scape" would not seem to have much significance beyond a specific societal anchorage, the metaphor of a landscape-of-marriages indicates the possibility of connections ("roads," "bridges") that are made possible by involving certain objects in the process. It is not only the connecting of different life worlds, styles, economies, partners and their families and social relations that take place but these connections also transform through time and space, making for a kaleidoscopic view of this ever-changing landscape. This calls for a combined approach to the agency of the people concerned as well as the agency of the linking devices that are employed.

A first element in this exploration of people and objects moving through this landscape is whether this is a world of increasing possibility of, or exposure to, the notion of agency and free choice. Western images of romantic love seem to be inciting couples in Africa to talk about their relationships increasingly in terms of "love" (Cole & Thomas 2009), contesting as outmoded the idea that marriages are arranged (by family elders) or that they follow certain structural preferences, such as the cross-cousin pattern. Some authors, such as Hirsch & Wardlow (2006), have argued that romantic courtship and companionate marriage appear as global success models, connecting couples in African situations to identity frames that have gained wide currency through the media, glossy magazines, and programs like the Oprah Winfrey Show.

From an anthropological perspective, in which studying marital relations has been so much part of the structuralist/structural-functionalist perspective on society, kinship, and family, this shift toward exploring agency in the context of marriage is not entirely satisfactory. There is not only the problem of who has access and the economic resources to aspire to these models in an African situation but also the significance of the commodities involved in the wedding process that should be analyzed in terms of what they *do*. What is the significance of these objects in the connections and relations that marriage creates?

This brings us to the linking agency of the objects involved in the process. Whereas marriage establishes relations between men and women and between families and communities, certain objects—wedding rings, wedding dresses, and wedding cards—can be seen to be used as linking devices. In studying lengthy marital processes in Botswana, marriage appears, in addition to being a form of relationship, also to have dimensions of connection with the capacity to connect, establish, and facilitate relations, much in the same way that a bridge is a linking device that enables the formation of diverse social relations. In understanding the difference between "relation" and "connection" (see also Feldman 2011 on the conflation between relations and connections), a difference in agency emerges as it allows for the possibility of exploring particular connecting devices that bring about relations between a local and a wider world and between different economies, social groups, forms of authority, and social expectations. By looking at how certain objects are centrally placed in weddings in Botswana as artifacts that have the capacity to connect, weddings can be studied as a context in which a nesting of connections takes place. While an extensive literature has emerged since Gell's (1998) work exploring the agency of objects (Appadurai 1986; Kopytoff 1986), it has been Knappett (2002: 101) who has demonstrated how the agency of an artifact is contingent upon the nature of its interconnections with other agents in a network. Nesting therefore means the ways in which marriage as a ritual of connectivity and relationality embodies a range of linking devices that demonstrates the capacity to link people to diverse social milieus, economic circumstances, and powers, and thus to influence their behavior.

Marriage in Molepolole

Weddings in the agricultural town of Molepolole are grand events, with the most visible aspect being a big marquee in the compound of the bride's parental home and the groom's parental home, which are lavishly decorated in bright colors. Here the couple will be seated and the important guests from both families will be invited to share the food and entertainment. These tents, which can seat hundreds of guests at nicely decorated tables, can be seen from afar and attract a great deal of attention. They are an ostentatious sign of the splendor of the wedding and indicate the wealth and status of the couple and their families. The bigger and more expensively decorated they are, the higher their status ambitions are. Though the hiring and decorating of the tents is expensive, which provides an interesting business opportunity for small companies all over the country, couples also need to be able to afford expensive dresses, gifts, food (including a cake), transport, wedding rings, invitation cards, photography, and all the other important commodities that make for a successful and prestigious wedding. Today's couples are feeling pressure from their parents (and less from their peers) concerning the enormous expenses they are faced with. "Parents" is used loosely here and does not explicitly refer to the biological father and/or mother who are among the many significant others and who, in fact, usually play only a minor role in the whole process. The central figure is the mother's brother (*malome*) and he is key in many decisions (Schapera 1940, 1950; Kuper 1982; Gulbrandsen 1986) as the *malome* from both parties conduct the premarital negotiations and assist in the bride-price payment.

There is a feeling on the part of couples that they need to negotiate to make their wedding as grand as possible. As one interlocutor put it: "You see, our parents expect us to live out their dreams; to do our wedding in a manner they themselves were never able to." The second source of concern is that parents also want to ensure that certain cultural obligations are met. These are the elements over which they feel they have full control. The bride-price payment (*lobola/bogadi*) is the most important aspect[1] and in Molepolole consists of the payment of eight head of cattle by the groom's family to the bride's family. This is a strict requirement without which the marriage will not be recognized locally. If the couple have had a child before getting married (which is very common), an additional head of cattle has to be paid as compensation for what is called "jumping the fence" (Molokomme 1991).

In the past, this *lobola/bogadi* represented the main element in the wedding process in terms of its cultural significance in cementing certain relationships of exchange and reciprocity, but also in terms of economic value (Schapera 1940; Comaroff & Roberts 1977). Nowadays, the relative significance of bride price has dwindled in economic terms and the relative value of the eight head of cattle has become marginal compared to all the other expenses involved in weddings. The following case of the couple S. is indicative of this: the groom is 32 years old and has a job in the IT sector and the bride is 31 years and has a job in the insurance business, both of them in Gaborone. The couple got to know one another in 2003 and had a baby soon afterward, and began the preparations for their marriage in 2007. Prior to marriage, the man's parents were required to meet the woman's parents because they needed to deal with the "accusation" that "your child has messed up our yard"; that is, engaged in a sexual relation with their daughter without any prior arrangement. Marriage negotiations could only start when a "repair" payment had been made to the woman's parents, namely, the *Tlhaga legora* (the fine for "jumping the fence"). Commonly, the baby's father is not forced to take care of the mother and the child although parents may try to enforce such arrangements with the help of the law. In practice, the success of the demands placed on the man amounts to little.

In all cases, a payment of one head of cattle has to be made to open the marriage negotiations. The so-called *patlo* (asking for marriage) that follows the bride-price negotiations usually sets the *lobola* at eight head of cattle. Of these, the groom commonly provides two, while the other are provided by his uncles, siblings, and even one from the *kgotla* to which he belongs. Yet, in this case he paid money for the eight head of cattle at a rate of BP. 1.200,- each. This was only the start of what the couple explained was a wedding that would cost them a total of BP 100,000, which is five times an average person's annual income in Molepolole. Celebrations were held at both the parental homes of the couple. Before the *patlo*, the woman had to be dressed and to have slept entirely in new clothing so the man provided her with new clothes, shoes, underwear, sheets, blankets, pillows, a lamp, umbrella, soap, and so on. Her mother was also given the customary blanket to put on her shoulders as a sign of maturity and authority.

For the days of the wedding, held late 2008, a tent had to be put up with all the decorations required to make it a wonderful, sumptuous sight, and food and drink for the hundreds of guests had to be bought and prepared, which included the slaughtering of one or two head of cattle. Dresses and shoes for both the bride and the groom were required as they changed their outfits several times during the wedding, as well as dresses for all the best men and the women who would be part of the ceremonies. Rings were purchased, cards were printed and circulated well in advance, and pastors were invited by the couple's parents to lead the prayers at the church wedding, which was a separate event. A traditional doctor had to be arranged to cleanse the places where the wedding was to be held, to prepare the food, and to guard the fires. He charged one head of cattle or the monetary equivalent (about BP 2,000). Music and a video man were arranged to provide entertainment and record all the stages in the wedding, and a cameraman was hired to make a photo shoot on the days of the wedding.

The couple were responsible for covering the huge expenses involved themselves. As with the cattle, relatives may step in to contribute but in this couple's case it became clear that most of the expense was going to fall on their shoulders so they decided to take out a bank loan. Since they both had paying jobs, they were able to borrow a considerable sum: the man took out a loan of BP 30,000 and the woman one of BP 10,000. The remaining BP 60,000 came from their savings and from small donations by relatives and friends, particularly on the groom's side. In many cases in which the *lobola/bogadi* is paid not in kind but in cash, the bride will paradoxically contribute to the bride-price payment from her own savings and/or a bank loan. Paying off these debts will take the couple years because of the high interest payments involved (about 20% annually), and repayment of their loan will probably take between five and ten years.

The couple's parents confirmed that certain elements in the marriage process have to take place to legitimize it in their eyes. The formal contacting by the groom's party of the family of the bride to indicate his intention of marrying one of their daughters is an important part. Payment of the bride price is an aspect they also clung to as something they saw as indispensable, as are other gifts too, as well as the traditional preparation of the food in the big cooking pots that is shared among all the guests on the wedding day.

Opinions differed regarding the glamorous nature of the wedding. Some of the parents said it was important that the wedding exude class and status. Other elders were, however, less convinced of this and took the position that young people today are doing things differently from the way they used to do it in the past and that the youth should decide the style of the wedding for themselves so as to be able to show off vis-à-vis their friends. Irrespective of the way in which the wedding was held, its location was as nonnegotiable as the bride-price payment. The marquees had to be placed in the garden in the ward of the parental homes. (Molepololele has over 70 different wards, each with a ward headman who is allowed to inspect weddings and supervise *lobola* payments.) Even if the couple, as in the case described here, does not live, work, or have a social life in Molepolole but in Gaborone, the wedding cannot be held in the city. In some cases, couples literally bring all the items they need from the city to the village. In terms of ritual, Gaborone is an "empty" space, a place devoid of symbolic meaning and therefore not a place where a sense of legitimation of a marriage can be produced. Yet, the wedding connects the couple's life world of Gaborone and its economy to that of Molepolole, a ritual connection that makes it possible to forge relations between the two families and between the couple concerned.

Generational Technologies of Connectivity:
The *Nama ya Tshiamo*

One element to consider when exploring the ritual connections that occur is what I propose calling "generational technologies of connectivity." What do they consist of in this particular situation? First, we need to remember that weddings today establish a specific ritual and economic connection between the city and the village, and between the life worlds of the emerging middle classes and a rural socioeconomic milieu. For the ethnic Bakwena group that established its paramount chieftaincy in Molepolole (Griffiths 1997), this is the place for the ritual negotiations surrounding the *patlo* when the customary elements of the presentation of the *lobola/bogadi* take place. If the bride is from Molepolole, this village becomes the primary ritual site for the forging of the marital bond. In many cases it also involves the participation of the traditional healer (*ngaka*) who, usually invited and paid by the groom's party, is required to spiritually protect and cleanse everything in the marriage exchange (gifts, rings, clothes, cows), ritually washing them and checking the position of the cooking fires and the food.

One specific item of food requires the special attention of these *dingaka*. In many instances, the couple is expected to eat a special meat called *Nama ya Tshiamo* that is prepared by the *ngaka*. The couple have to take the "first bite" (*molomo*) jointly before anyone else can enjoy the food because the meat is an object of connection. It is a linking device that has an agency of its own in terms of making possible a variety of relations, each of which is considered quintessential for the marriage, the success of the bond, relations between the two families, and relations with the ancestors. What is this connecting device with its remarkable agency?

Nama ya Tshiamo (literally, the meat of righteousness/fullness/well-being) is cut when the first beast is slaughtered at the parental compound; Figure 7.1 shows this special piece of meat which here is held by the traditional healer (*ngaka*) who will play and important role in preparing it. Usually cut by the bride's mother's brother (*malome*) following instructions from a *ngaka*, it is a piece of meat that allows for the literal consummation of the marriage as it needs to be eaten by the bride and the groom in the company of their most important relatives. After the meat has been cut from the freshly slaughtered animal, the *malome* takes it to the house where the *ngaka* will treat it with a range of medicines indoors and out of sight of others in the compound. The *malome* will then roast it over a special fire and when it is ready for consumption and has been cut up, the couple will go into the house together to eat it.[2] The beast is usually slaughtered on the Friday morning and the *Nama ya Tshiamo* is consumed the same afternoon. All the preparations, including the cooking, can then continue so that the festivities can start on Saturday.

Different meanings are attributed to this piece of meat, its preparation, and its consumption. Many people, as well as the *dingaka* I interviewed, explained that the meat, the medicines inside it, and its joint consumption strengthen the bond between the couple. Its working is that it "brings the couple together," many explained. The medicines protect the couple against jealousy, spite, and envy that could cause them to split, be divided amongst themselves, fight, or come into conflict with their social environment. It joins the couple spiritually and enables their "blood" (*madi*) to come together fruitfully so the meat and its joint eating are seen as being conducive to

Figure 7.1 The ngaka (traditional doctor) holding the *Nama ya Tshiamo*, freshly cut from the slaughtered cow. (Molepolole, November 2010, picture by author).

successful reproduction. The meat is cut from the part of the slaughtered cow where the hind leg joins the body of the beast as it is believed that this is the place where different streams of blood—one vein running from the head of the animal to the leg and one vein running from the leg up toward the body—actually meet, that is, the place of a literal meeting of different bloods. This binding-in-the-flesh requires an unbinding at the moment of death of one of the couple. When a couple has taken the *Nama ya Tshiamo* at the start of their relationship, special mourning rituals have to be performed to undo the powers embedded in this linking. These cleansing funereal rituals are known as *goribama* and if they are not performed properly, this will lead to a specific kind of spiritual affliction known as *boswagadi* (Kealotswe 2007).

As an artifact of linking, the *Nama ya Tshiamo* needs to be treated with extreme care. In terms of its capacity for connecting, our point of departure for further analysis of this agency is the question: What is it linking?

As a linking device, the *Nama ya Tshiamo* makes blood relations possible between the husband and wife and between the couple and their offspring (Klaits 2005).[3] Some older men explained that greater emphasis was placed

on *Nama ya Tshiamo* in the 1920s and 1930s when many men were starting to migrate to the South African mines and industrial complexes in search
of labor. The meat ensured a prolonged and extended marital connection in
times when a marital relationship was hard to maintain because of distance
and the long absence of married men (Comaroff & Roberts 1977; Brown
1983; Townsend 1997). One specific "activating" agent that was and that still
can be added to the meat in preparation for the couple to eat was and is the
thotse, also known as "the seed." Consuming this while eating the meat means
that the husband is required to come home before the year's ending and have
intercourse with his wife. If the couple does not do so, the expectation and
fear is that any extramarital relation can cause serious harm to their health
and that of their offspring. However, if the *thotse* is honored and a reconsum-
mation of the marital relation has indeed taken place at the appropriate time,
the engagement with these extramarital relations was and is not expected to
cause such threats. The eating of the meat therefore makes the safeguarding
of the marital bond Janus-faced as being involved in concubinate relationships (the so-called *inyatsi* or "small houses") will not threaten the continuity
of the marital bond. In such extramarital relations, this special meat will not
be eaten, thus ensuring the exclusivity of the marital bond. While the efficacy
of the *thotse* is believed to have been of more relevance in the years when
men were involved in long-distance labor migration to the mines, helping to
ensure a periodic return of the men to their wives, many in Molepolole still
referred to its significance for present-day relations.

The *Nama ya Tshiamo* also expresses the temporality of the life course in
the way it links the before and the after of the marital relationship. It links
its start and its ending, and thereby produces a teleology, a purposefulness of
its (protective) coverage over a period of time. In addition, it links the visible
with the invisible world of spiritual powers, which are ambiguous and can be
good or bad and afflictive. It ensures a proper relationship with these powers and provides spiritual protection to ward off evil powers of jealousy and
witchcraft. The meat should not be cooked thoroughly: it should be half-raw
and red when eaten as the taste of the blood of the slaughtered animal signifies
the blood that the couple will share and bring together through intercourse,
thereby continuing a protected ancestral line of procreation.

The meat provides a profound linkage with the authority of the elders and
emplacement in hierarchical kinship relations. Eating the meat demonstrates
submission to these powers and signals the fact that, as a young couple, one
is prepared to accept and acknowledge this structure. In this sense it links the
agency of the couple to this structure to the extent of literally "taking in" all
that the elders' authority presents them with. Submitting to eating is in this
sense a powerful marker of how one places one's agency, one's body, and one's

metabolism in a context of such a structure of power. It is a relationality of which the presence of the couple's children is a further signifier.

The consuming of the meat connects with the production of the meat. Being in a bovine culture in Molepolole, the production and keeping of cattle and the selling of meat, milk, and hides are important elements of production and social status. While many people hold jobs and have land (*masimo*) on which they cultivate crops for the market and their own consumption, cattle herding and breeding are a sure sign of (the man's) economic potential and of a strong patriarchal kinship structure. While there is patriarchal control of these resources, the eating of meat signals a relationship with a moral economy of production: it is the eating of *beef* and nothing else. The couple eats from what the men produce and this suggests the line of dependency that women and families find themselves under. The wife marries into her husband's family and her productive powers are placed in their service.

The eating of the meat links families. It is not only the couple that eats this meat but the *malome* on both sides also take a bite, as can their wives and some other older relatives. The couple takes the first bite (*molomo*) but this prepares the way for the families to be united in sharing the meat and in the potentialities of future procreation. Much of the practice of *Nama ya Tshiamo* is based on the notion that instead of individuals it is actually families that marry, thereby indicating a specific connection of the world of individuality, individual desires, and prospects with that of a wider communal understanding of the significance of relations and reproduction.

Molepolole represents in many ways a polysemic ritual site where connections are not only multilayered but above all nested: in marriage there is a linking within linking. A connecting device such as the *Nama ya Tshiamo* is, in the way it makes possible a diverse set of relations (between people, different worlds and economies), included in a higher level of connectivity that the marriage represents. Generational relations dictate much of this nested connectivity and the ways in which an object of linking is capable of doing that linking cannot be removed from the ways in which the older generation forges a network between Molepolole and the nearby city, between the economy of the agro-town and that of the urban, and between the lives of the young emergent urban groups with their forms of paid employment and their lifestyles but with their representational obligations in the rural area as dictated by the elders. The ritual economic connection between the city and the village involves the exchange of luxury items and commodities, and a large part of the expenses take place within the village. In a sense, the village benefits from the urban economy where the young people's money and incomes contribute to the reciprocities (food, cattle, gifts) that take place in the village. Interestingly, the young people that I interviewed who were living in Gaborone but had got married in Molepolole felt afterward that they were "left to themselves," "left without support," and "saw nobody any longer." These quotes can be interpreted as indicating that couples enter a new stage in their lives following marriage, one in which they are responsible for themselves in economic terms and can no longer rely on their parents. However, they also indicated a feeling of having invested in family relations as part of this urban-rural ritual connection, while little had emerged in return.

The bigger picture here is one of a more overarching form of connectivity (between these life worlds) that comprises and envelops forms of connectivity that specific linking devices represent. Yet while this nesting seems to present a specific order of the way in which particular objects serve levels of connectivity, is this running uncontested from the perspective of the young themselves?

When Connections Contest Relations

One Friday in November 2010, Ms. S., the bride of Mr. K., caused turmoil among her and her husband-to-be's relatives and friends who were preparing their wedding festivities: she ran away and took a bus back to Gaborone where she disappeared without trace. The embarrassment and confusion were considerable. The *Nama ya Tshiamo* had just been cut, the *ngaka* had been called to doctor it with his medicines, and the special meat had been put on the fire to cook. And now the bride had disappeared. I had been invited to attend the slaughtering of the beast and the subsequent preparations for the wedding at the bride's home and a few days before the actual event I had talked to her about these arrangements. While elderly women were in her presence, she had told me: "We are Christians, we do not believe in these things. There will be no *Nama ya Tshiamo* as me and my husband are Christians. Our families understand." Being interested to see what the wedding preparations amounted to, I had come on the Friday to the *kgotla* in Molepolole where the beasts were being slaughtered and where many people had gathered to help in this dirty work (which usually means they will receive a share of the meat in exchange). And I had witnessed the cutting and preparation of the special meat, only to find the *malome* of the wife and the *malome* of the husband being at complete loss as to how to continue now that a "first bite" was not going to take place. The young woman had caused a disaster by her absence: preparations had to stop, fires were extinguished, and the men discussed what should be done as the beasts had been slaughtered and skinned and the meat should not sit out in the heat of the Botswana sun for too long.

The *Nama ya Tshiamo* was causing a *disconnection* instead of a connection. Now the elders were placed in a difficult position and needed to do some damage limitation in the face of this embarrassment. The young woman had clearly, in their minds, not acted as an adult or a mature person although marriage would have conferred on her new responsibilities.

Marriage, after all, is *the* step toward adulthood. Having fulfilled the payment of *lobola* and gone through all the traditional obligations and rounds of advice and organized the festivities establishes both partners as adults and gives them a voice and visibility. There is no better moment for an exercise in mature adult power than the marital proceedings themselves. Every marital arrangement or wedding that I have attended has been marked by conflicts between the families involved that created the deliberate arenas for the adults, that is, the married male or female members of the family, to raise their voice and be seen and heard. Even the smallest detail could lead to elaborate discussions between the adult family members, involving fines in most cases for the party of the groom or serious delays in the entire wedding process. In one case, for example, the payment of the eight head of cattle resulted in consternation following inspection by the receiving party, as one of the bulls had not been castrated. This was considered a serious offense and the adult men on the receiving end went into a meeting that lasted a whole morning to discuss the proper way to respond to this insult and decide on the level of fine that the other party should pay. (It was decided that they had to pay BP 300.) Part of the generational technology here is the skill and competence to create conflict, to make one's voice heard and to negotiate the process of picking a fight.

Having gone through the marital process allows the husband and wife to take part in (extended) family affairs and in decision-making processes that concern rights over property, children, inheritance, and the marital affairs of others in the family. While the social meaning of adulthood to many involves the notion of taking on responsibilities concerning one's own financial matters and becoming independent and self-reliant, marriage connects a couple to a cultural-legal position that confers special powers on them. Having acquired the social status to claim rights and give voice to them is reserved for those who have passed the initiation of marriage successfully, although these rights to a voice are unequally shared between men and women.[4]

Playing havoc is reserved for the elderly, particularly for adult men. Yet in this case havoc was created by a young woman who had voted with her feet. How can we analyze the reasons for her actions and for refusing the *Nama ya Tshiamo* and the connections that this object engenders?

First of all, it is important to understand that marriage and its arrangements produce a space in which the young can try to underscore a sense of their independence from the older generation. Their spending power and level of education are usually much greater than those of their parents, and their access and exposure to urban styles are similarly more extensive. While the parents may exercise control over certain key elements in the marriage process as part of their generational technology, the younger generation should not be seen as passive players, as being the recipients or even the victims of this generational technology (Alber *et al.* 2008). They are partners in the production of this generational technology on their own account and through their own means.

The question of how and which social technologies come into play in producing a generation are related to urban styles of consumerism, global ideas of romantic relationships, partner choice, as well luxury commodities, which the older generation have little access to. Increasingly, stressing marriage as the desire for a particular (romantic) relationship seems to be part of their social technology in creating a sense of "generation." Ms. S. had been exposed to this new and modern life styling in Gaborone where she had been employed in a fashionable music/video production company that sold to the middle-class urban market. She was comfortable surrounded by modern Yuppie flamboyance, which was accessible to her thanks to her reasonable income. Her husband-to-be, a 43-year-old, well-trained male nurse, was receiving in-service training in London and Oxford and also enjoyed a comfortable income. Their lifestyle with its international travel bordered on the cosmopolitan, a style they were eager to transmit in their wedding as well.

Christianity's presence in Molepolole since the time of Livingstone has left its mark on society and generational relations. It has a long record of promoting monogamy in spite of the Bakwena chieftaincy's initial reluctance to abandon polygamy. Other than this, the established mission-based denominations did not intervene much in marital arrangements until the arrival of Pentecostal Charismatic groups in Molepolole in the mid-1970s. They were eager to intervene much more in marital arrangements than the established denominations of the United Congregational Church of Southern Africa (UCCSA); the Anglican or the Roman Catholic Church had been inclined to and impressed on their members that no traditional doctor should play a role in any aspect of their wedding preparations. But in particular they turned the practice of the joint eating of the *Nama ya Tshiamo* into a bone of contention. As Ps. S. of the popular Apostolic Faith Mission Church recounted, since establishing the church in Molepolole they had been very concerned about this "demonic' practice" and had to "explain" to people in no uncertain terms that this practice was unbiblical and could invoke dark powers and keep people in bondage.

While some of the mainline churches would certainly have similar concerns about the spiritual dimensions of the eating of this meat, they found it much harder to intervene and supervise their members. The Pentecostal pastors put in place an elaborate system of premarital counseling through which the couple to be married would be guided throughout their wedding preparations. Such Pentecostal counseling stands diametrically opposed to the

formal moment of traditional "counseling" by the elders that takes place separately for men and women during the marriage ceremony. In Pentecostal
marital counseling there is emotional and social-cultural preparation for married life and they might take offense at traditional sessions because of the
moralities they would find anti-Christian. Many young Pentecostals made
it clear to me that they saw traditional advice giving as being "empty" as
"we don't get anything from that talking." What they got from counseling
by the elders at the marriage ceremony was generally perceived as meaning-
less. Words such as "a man is like an axe; everybody can pick it and use it"
or "a man is like a dog, he goes here and there" appear to create a sense of
license for extramarital affairs that run counter to their Christian perspectives.
In this sense, the ritual connection with Molepolole and the cultural obligations (such as going through traditional counseling) can be fraught with issues
of generational power and tensions of an ideological nature.

Were Ms. S. and Mr. K. Pentecostals therefore forced to reject the *Nama
ya Tshiamo* and disregard the advice given to them by their elders? Neither
of them was: Ms. S. belonged to the mainline UCCSA and Mr. K. did not
profess any church membership at all nor did he come from a Christian background. Pentecostal talk about the spiritual nature of the *Nama ya Thsiamo*
had surely, however, reached them in the cultural milieu of the city. And yet
they decided differently. Ms. S. refused to eat to meat but Mr. K. did do so
and it was rumored that the mother of Ms. S. stepped in as her daughter's
replacement so as to take the first bite late, thereby ensuring that the wedding
preparations were able to continue.

What the *Nama ya Tshiamo* did in this case was to demonstrate a fracture
in the power of the elderly and a level of independence on the part of the
young. The meat not only linked partners, families, economies, and spiri-
tualities but became a catalyst in the delinking of these as well. Ms. S. and
Mr. K. were now not spiritually linked in the sense that specific mourning
cleansing rituals would be required if either of them died. And the protective capacities of the medicines to enable sustenance of the marital bond were
not consumed. While the meat acts as a linking device in many ways, in
this case it also demonstrates its flip side of delinking along the fault lines
of generational relations and, in this case, more for a young woman than for
a man.

Conclusion: Marriage as a Relation and a Connection

Studying marriage allows for two different but closely connected perspectives to emerge that shed light on the specific social technologies at play
and produce this bond. Taking a view of marriage from the perspective

of "relations" produces an analysis that focuses on young people's pursuit of transculturalization with an appropriation of Western or global images of romance, middle-class styles, and consumerism. Here agency seems to be colored by the youth's desire to construct a sense of generation and generational power for themselves: the social technologies they pursue are made possible by higher incomes, greater spending power, and access to global images of white weddings. In this perspective questions relate to how new notions of love, romance, and affection may shape or refashion the nature of the relationship.

Looking at the same marriage rituals from the perspective of "connections" shows a different set of questions that analytically need to be separated from those that govern relations. It is too easy to perceive of this difference as agency versus structure because both perspectives offer insight into how structural demands are playing out against reflexive agency, against how individuals take decisions on the basis of their perception of structural conditions (de Bruijn et al. 2007).

The connections perspective of marriage in Botswana enables us to see how marriage is creating specific structuralizing bonds between generations, places, and different life worlds. Investigating ritual (such as marriage) as a form of connectivity—in this case between different lifestyles and traditions, between rural and the urban economies, and between the generations of the young and their elders—allows for an exploration of the dimensions of power and hierarchy that are embedded in it. While the young want to negotiate space for marriage in terms of its romantic ideals, they also need to negotiate connectivities in terms of becoming recognized, becoming adults, and securing their rights.

They also need to negotiate the fact that there are linking devices within the linking that the marriage process includes. In this nesting of connections they need to negotiate the agency of certain linking devices, one of which was described as being the *Nama ya Tshiamo*. Again, the analytical distinction between "relations" and "connections" applies in the understanding of this negotiation. In the way the *Nama ya Tshiamo* acts as a linking device it facilitates a range of social relations, changing their quality and significance. And as such, it also appears to have the capacity to turn relations sour, making them tense and fraught with anger or embarrassment. In this way, linking devices (like the "bridge") take on a social life of their own, much as has been described for the social life of any other artifact in human relations (Appadurai 1986; Kopytoff 1986; Gell 1998; Knappett 2002). But as the linking device remains nested in other human-inspired connections, their agency in this social life cannot be separated from the human agents themselves that put in place—or remove—the linking devices.

It is also interesting to note how difficult it is to reconcile these dimensions. While this modern struggle can also be taken to inform anthropology on how to analytically separate a perspective of relations for one of connections and argue they are not the same—as also Feldman (2011) is succinctly arguing—in the second instance both perspectives need to be brought together again to make sense of reality. A complete picture of marriage arrangements in Molepolole only emerges if the two perspectives are combined, while in the final analysis the one can never take precedence over the other.

Another angle from which it is important to perceive marriage from a connections perspective is that connections always introduce disconnections at the same time and therefore the ways in which marriage produces ritual disconnections need to be taken into account too (see Bochow, this volume). The empirical material presented here discusses disconnection—married people now being on their own, taking their own decisions, daughters becoming detached from their mothers/families as they are physically and literally required to move from their parental home to that of their husbands, and so forth. When talking to couples, the need to distance themselves from their parents, establish their own independent households, and ensure that there is no interference from their in-laws is indeed a difficult experience of disconnection while keeping relations intact as far as possible. And perhaps this experience is also difficult for the parents, thus raising the question as to whether the wedding is a connecting symbol that "covers" the subjective experience of "disconnection" as well.

By asking this question we are in fact answering it. Apparently, perceiving marriage as a (dis)connection allows us also to see certain particular messages that the ritual is communicating. This is a polyphone message of how the ritual communicates the meaning of urban economies, global styles, as well as personal experiences of adulthood and relational attachments. The advantage of opening up a connections perspective on marriage is therefore that it allows us to see different layers of meaning than only those relating to the more obvious messages of reciprocity, consumerism, class, status, or romantic ideals.

Notes

1. There is an extensive anthropological literature on the *lobolab/ogadi* bride-price payments in cattle in Botswana, starting with the early work of Schapera (1940, 1950) and later studies by Comaroff & Roberts (1977), Kuper (1982), Solway (1990) and, for Molepolole, Griffiths (1997).
2. Variations occur in the process of consuming the *nama ya tshiamo* as couples and their important others may eat the meat with their hands, while in other cases the

malome takes a small sharp stick, pierces the meat onto the top of the stick, and puts the meat into the mouths of the bride and groom so that they do not touch it. Sometimes the couple's children consume part of the meat too.

3. Klaits also points at the fact that ideas concerning the sharing of blood between couples is significant in the people's understanding of HIV/AIDS and the way in which the disease has been spreading.

4. See Griffiths (1997) on the legal-anthropological understanding of this point in marital cases presented to the *kgotla*. Although the government is against the practice, property grabbing still takes place when a man dies and his relationship with a particular woman has never been fully recognized at a cultural-judicial level. This is one of the main reasons why older couples who have been living together for many years still want to get married. Childbirth does not bring adulthood so formalizing their bond through marriage remains important.

References

Alber, E., S. van der Geest & S. Reynolds Whyte (eds.) (2008) *Generations in Africa: Connections and Conflicts*. Berlin: Lit Verlag.

Appadurai, A. (1986) "Introduction: Commodities and the Politics of Value," in: A. Appadurai (ed.), *The Social Life of Things: Commodities in Cultural Perspective*. Cambridge: Cambridge University Press, pp. 3–63.

Bochow, A. (2008) "Syncretism and Antisyncretism in Marriage Patterns in Kumasi: The Role of Pentecostal Charismatic Churches," in: A. Adogame, M. Echtler & U. Viercke (eds.), *Unpacking the New: Critical Perspectives on Cultural Syncretization in Africa and Beyond*. Hamburg: LIT, pp. 239–266.

Bochow, A. (2010) "Intimität und Sexualität vor der Ehe. Gespräche über Ungesagtes in Kumasi und Endwa, Ghana," Ph.D. Thesis, Bayreuth University. Hamburg: LIT Verlag.

Brown, B. (1983) "The Impact of Male Labour Migration on Women in Botswana," *African Affairs* 28(328): 367–388.

Bruijn, M.de, R. van Dijk & J-B. Gewald (eds.) (2007 *Strength beyond Structure: Social and Historical Trajectories of Agency in Africa*. Leiden: Brill.

Cole, J. & L. M. Thomas (eds.) (2009) *Love in Africa*. Chicago: University of Chicago Press.

Comaroff, J. L. & S. Roberts (1977) "Marriage and Extra-Marital Sexuality: The Dialectics of Legal Change among the Kgatla," *Journal of African Law* (1): 99–123.

Constable, N. (2009) "The Commodification of Intimacy: Marriage, Sex, and Reproductive Labor," *Annual Review of Anthropology* (38): 49–64.

Cornwall, A. (2002) "Spending Power. Love, Money, and the Reconfiguration of Gender relations in Ado-Odo, Southwestern Nigeria," *American Ethnologist* 29(4): 963–980.

Feldman, G. (2011) "If Ethnography Is More than Participant-Observation, then relations are More than Connections. The Case for Nonlocal Ethnography in a World of Apparatuses," *Anthropological Theory* 11(4): 375–395.

Gell. A. (1998) *Arts and Agency: An Anthropological Theory.* Oxford: Oxford University Press.

Griffiths, A. M. O. (1997) *In the Shadow of Marriage: Gender and Justice in an African Community.* Chicago & London: University of Chicago Press.

Gulbrandsen, O. (1986) "To Marry—Or Not to Marry. Marital Strategies and Sexual Relations in Tswana Society," *Ethnos* 15(1 & 2): 7–28.

Hirsch, J. S. & H. Wardlow (2006) *Modern Loves. The Anthropology of Romantic Courtship and Companionate Marriage.* Ann Arbor: University of Michigan Press.

Johnson-Hanks, J. (2007) "Women on the Market. Marriage, Consumption and the Internet in Urban Cameroon," *American Ethnologist* 34(4): 642–658.

Kealotswe, O. N. (2007) "The Church and HIV/AIDS in Africa: An Overview of Ethical Considerations," in: J. Amanze, O. N. Kealotswe & F. Nkomazana (eds.), *Christian Ethics and HIV/AIDS in Africa.* Gaborone: Bay Publishing, pp. 14–27.

Klaits, F. (2005) "The Widow in Blue: Blood and the Morality of Remembering during Botswana's Time of AIDS," *Africa* 75(1): 46–62.

Knappett, C. (2002) "Photographs, Skeuomorphs and Marionettes," *Journal of Material Culture* (7): 97–117.

Kopytoff, I. (1986) "The Cultural Biography of Things: Commoditization as Process," in: A. Appadurai (ed.), *The Social Life of Things: Commodities in Cultural Perspective.* Cambridge: Cambridge University Press, pp. 64–91.

Kuper, A. (1982) *Wives for Cattle. Bridewealth and Marriage in Southern Africa.* London: Routledge & Kegan Paul.

Mann, K. (1985) *Marrying Well. Marriage, Status and Social Change among the Educated Elite in Colonial Lagos.* Cambridge: Cambridge University Press.

Masquelier, A. (2004) "How is a Girl to Marry without a Bed? Weddings, Wealth and Women's Value in an Islamic Town of Niger," in: W. van Binsbergen & R. van Dijk (eds.), *Situating Globality. African Agency in the Appropriation of Global Culture.* Leiden: Brill, pp. 220–256.

Masquelier, A. (2005) "The Scorpion's Sting. Youth, Marriage and the Struggle for Social Maturity in Niger," *Africa* (11): 59–83.

Molokomme, A. (1991) *Children of the Fence. The Maintenance of Extra-marital Children under Law and Practice in Botswana.* Leiden: African Studies Centre.

Pauli, J. (2009) "(Re-)Producing an Elite. Fertility, Marriage and Economic Change in Northwest Namibia," in: C. Greiner & W. Kokot (eds.), *Networks, Resources and Economic Action.* Berlin: Reimer, pp. 303–325.

Schapera, I. (1940) *Married Life in an African Tribe.* London: Faber & Faber.

Schapera, I. (1950) "Kinship and Marriage among the Tswana," in: A. R. Radcliff-Brown & D. Forde (eds.), *African Systems of Kinship and Marriage.* London: Oxford University Press, pp. 140–165.

Solway, J. S. (1990) "Affines and Spouses, Friends and Lovers: The Passing of Polygyny in Botswana," *Journal of Anthropological Research* 46(1): 41–66.

Soothill, J. E. (2007) *Gender, Social Change and Spiritual Power. Charismatic Christianity in Ghana.* Leiden: Brill.

Townsend, N. W. (1997) "Men, Migration, and Households in Botswana: An Exploration of Connections over Time and Space," *Journal of Southern African Studies* 23(3): 405–420.

van Dijk, R. (2004) "Negotiating Marriage: Questions of Morality and Legitimacy in Ghanaian Pentecostal Diaspora," *Journal of Religion in Africa* 34(4): 438–468.

CHAPTER 8

Connecting Communities and Business: Public-Private Partnerships as the Panacea for Land Reform in Limpopo Province, South Africa

Marja Spierenburg, Ben Cousins, Angélique Bos, and Lubabalo Ntsholo

Introduction

Scattered among large-scale citrus orchards and game farms in Limpopo Province lie the densely populated former homelands of Venda, Gazankulu, and Lebowa. With few possibilities for development in these barren areas, many communities have lodged claims for the restitution of land from which they were evicted under apartheid. Limpopo Province has a relatively high number of land claims compared to other provinces but has been slow in responding to them (Ramutsindela 2007). Influenced by reports documenting the "failure" of restitution projects elsewhere in South Africa (Hall 2004; Lahiff 2008), the provincial authorities feared that restitution would damage the province's economy, which is largely based on export-oriented agriculture and tourism (www.limpopo.gov.za; Pauw 2005), and has sought a way to do justice to the land claimants and maintain the productivity of the farms that were to be transferred. To this end, the provincial authorities obliged restitution beneficiaries to enter into a partnership with a private for-profit organization and develop a business plan to continue running their farm as a large-scale commercial enterprise (Fraser 2006, 2007). The aim of these strategic partnerships is to prepare land-reform beneficiaries to take over commercial operations after a period of (usually) 10–15 years, during which time the private-sector partner is responsible for skills transfer. These connections between communities of beneficiaries and the (rural) business sector—so-called strategic partnerships—are supposed to integrate the former into the modern commercial agricultural sector and connect them with global agricultural markets.

Since South Africa's transition to democracy in 1994, its agricultural sector has become one of the most liberalized in the world (Atkinson 2007). Terreblance (2002) describes how, during the discussions preceding this transition, African National Congress (ANC) cadres were "prepared" by British and American economists and were taught the virtues of neo-liberalization. These lessons, which were apparently taken very seriously in developing post-1994 agricultural policies, have managed to transform the sector from being heavily subsidized during apartheid to one of the least subsidized in the world. The land-reform program, instituted to redress the injustices of a highly racially skewed distribution of land inherited from the apartheid days, has increasingly moved in the direction of a more market-oriented and commercially oriented policy (Lahiff 2003; Hall 2004). The strategy adopted by the Limpopo Provincial Government of forcing land-reform beneficiaries into partnerships fits well with a neo-liberal approach to agriculture and land reform.

These strategic partnerships are a form of public-private partnership (PPP), established by private profit organizations and local government institutions representing the land-reform beneficiaries, with government offering subsidies to develop joint business ventures (these are about the only subsidies left in agriculture in South Africa). PPPs are increasingly popular in the field of international development cooperation and sustainable development. Although they are not a new phenomenon (Linder 1999), their popularity in policy circles has been steadily increasing since the late 1980s (Entwistle & Martin 2005) to the point where their promotion seems to have become a dominant "development narrative" (cf. Roe 1991, 1995). PPPs are being promoted as the most logical solution to a variety of service-delivery and development problems, and are often presented as technical and politically neutral solutions (cf. Ferguson 1990). Nevertheless, the promotion and development of PPPs have a distinct ideological background and flavor (Linder 1999; Entwistle & Martin 2005). The present popularity of PPPs came in the wake of a wave of privatization of government institutions by conservative governments in Europe and the United States—notably the Reagan administration and the Thatcher government—in the 1980s. The idea of the need for the privatization of government services was exported to developing countries through the many Structural Adjustment Programs enforced by the International Monetary Fund (IMF) and supported by the World Bank.

These were associated with increasing economic hardship and decreasing levels of service delivery to the poor in developing countries. In the 1990s, the focus shifted from privatization to PPPs, which were considered a "softer" version of the same process (Entwistle & Martin 2005) that would have less dramatic social consequences and therefore be more palatable to the general public. New Labour in the United Kingdom subsequently stressed the partnership idea in PPPs and the influence it was supposed to accord not only to the corporate sector but also to civil-society organizations (*Ibid.*).

There is a growing body of literature on PPPs, yet a lot is quite prescriptive (Brinkerhoff 2002) and written from a public-administration or management perspective, focusing on the core characteristics of PPPs and providing recommendations concerning the setting up of such a construction. Rarely does this kind of literature address the ideological underpinnings of the promotion of PPPs, nor does it question the concept of partnerships and the inherent power relations within PPPs. In much of the literature it is furthermore suggested that the public sector can learn more from the private sector in terms of efficiency, orientation toward results, and flexibility than the other way around (Brunsson & Sahlin-Anderson 2000; Batley 2004). There is only limited attention to how the concept of PPPs is translated and/or transformed as it travels across the globe and across sectors (cf. Monaci & Caselli 2005) and what the relative influence of the different partners involved is (Brinkerhoff 2002).

This chapter analyzes how the popular "panacea" for economic development and service delivery of PPPs is connected to and translated in the land-restitution program being implemented by Limpopo Province. It considers the inherent paradoxes of PPPs in land reform in Limpopo Province, which on the one hand supposedly foster the transformation of the agricultural sector into a more equitable one, while, on the other hand, they are an expression of the state's promotion of the continuation of large-scale commercial farming. The partnerships may serve as bridges for formerly dispossessed communities to (re)enter South Africa's rural economy. However, they are also meant to ensure the continued operation of large-scale commercial farming, with a focus on large-scale, export-oriented citrus plantations in the case of Limpopo Province. Communities are being discouraged from moving back onto their reclaimed land and are supposed to operate the plantations jointly with their new business partners from their bases in the former homelands. This entails the risk that no real transformation takes place on the plantations themselves and that it is still business as usual, with the same unequal labor relations, especially for those laborers who do not belong to the claimant communities.

This highlights an often-neglected aspect in the literature on PPPs, namely, the issue of power relations. These partnerships are all too frequently presented as universally applicable organizational technologies that automatically lead to a win-win situation (Brinkerhoff 2002). This chapter, however, analyzes the possible impact of power relations on the distribution of costs, risks, and benefits among the partners involved in strategic partnerships in Limpopo Province. Attention will also be paid to the heterogeneous character of communities of land-reform beneficiaries and the distribution of costs, risks, and benefits within these groups. In other words, the transformative potential (or lack thereof) of the connections between land-reform beneficiaries and private-sector organizations will be considered, as will the mechanisms of inclusion and exclusion related to these connections.

Public-Private Partnerships: The Traveling of Powerful Ideas

Proponents of PPPs present them as a new generation of management and governance tools that are "especially suited to the contemporary economic and political imperatives for efficiency and quality" (Linder 1999: 35). In the 1970s, PPPs were popular in the United States in fostering the development of inner cities (*Ibid.*) but their contribution to development was at best mixed (Stephenson 1991; Linder 1999). One has to wonder therefore why they resurged in the late 1980s.

The United Kingdom and the United States saw the advent of conservative administrations in the 1980s that were bent on reducing state expenditure and increasing the role of market forces. A wave of privatizations of public services and corporations followed (Starr 1988; Linder 1999; Entwistle & Martin 2006). The fall of the Berlin Wall at the end of the decade strengthened belief in the appropriateness of this approach. Through the international monetary institutions such as the IMF and the World Bank (the Washington Consensus), the approach was exported to developing countries that could no longer rely on the Eastern bloc, and ultimately even to the Eastern bloc itself (Wedel 2003). However, the restructuring of the state and the privatizations of governmental departments and services, introduced through Structural Adjustment Programs, led to social unrest as jobs were lost and government subsidies cut, rendering services such as education and healthcare less accessible to the poor. In the mid-1990s, when the conservative parties in the United Kingdom and the United States lost the national elections, the new governments did not abandon their market-driven approach but opted for a neo-liberal market-driven approach instead of a neo-conservative one. The focus shifted from privatizations to the promotion of PPPs, which were seen as "softer" versions of privatization that would be more palatable to the general public (Linder 1999).

The main idea behind this was that by reducing the role of the state, forging stronger connections between the state and the private (for profit and nonprofit) sector, and allowing a greater role for the market, service delivery to citizens would become more efficient and flexible (Linder 1999; Entwistle & Martin 2006). The public sector was also subjected to reforms based on the notion that the public sector should perform in a more businesslike manner, become more efficient in service delivery, and respond to the market. These reforms, often referred to as New Public Management (NPM), took an abstracted private for-profit organization model as its point of reference, according to Brunsson & Sahlin-Andersson (2000). Proponents of NPM portray the public sector as inefficient and slow in responding to changes in society. Hence, the focus in many cases is mainly on what the public sector can learn from the private for-profit sector, and little attention is paid to what the latter can learn from the public sector. Critics warn that, as a result, issues such as accountability and democratic control over the public sector are being ignored (*Ibid.*). They question the necessity of the directionality of the flow of ideas and principles and note that the public's perceptions of and demands from the private for-profit sector are also changing, and that when it comes to, for instance, Corporate Social Responsibility (CSR), the private for-profit sector might learn from the public and nonprofit sector too.

Yet the idea of the need to learn from the private for-profit sector appears to be powerful and has been extended to the private nonprofit sector, especially through the New Policy Agenda (NPA). This stresses the need for more contract-like relationships between the private nonprofit sector and donors. Donors of the private nonprofit sector are increasingly demanding that the latter demonstrate their efficiency by measuring their performance and detailed accounting of their expenses. Edwards & Hulme (1996) warned that this could result in nonprofit organizations becoming more accountable to donors rather than to the people that could benefit from their activities. They also warn that many of the lessons that the nonprofit sector learned the hard way are being ignored now that the sector is becoming more businesslike. These include, for instance, the idea that participation, empowerment, and local ownership of socioeconomic development processes is crucial but that participation and development are long-term processes and notoriously difficult to measure, monitor, and predict. Monaci & Caselli (2005) are less pessimistic, arguing that what they term "market isomorphism" does not occur through a process of diffusing ideas from the profit sector to the nonprofit sector but through a process of translation. Certain ideas and processes from the profit sector, which actors from the nonprofit sector deem valuable as well, are not simply copied but translated in such a way that they apply and are useful to the nonprofit sector. Nevertheless, they warn that in some cases, governments and/or donors may impose certain principles from the profit sector on the nonprofit sector.

Let us now turn to the issue of power relations. In much of the prescriptive literature on PPPs, a great deal of emphasis is placed on the need for good communication between the different parties involved (Brinkerhoff 2002). This assumes that "miscommunication" is often inadvertent and not part of a deliberate strategy to gain the upper hand. What is ignored in much of this literature is that not all the partners in a partnership are equal. Differential access to information may play a role but, as mentioned above, dependence on funding may also influence power relations (Edwards & Hulme 1996; Klijn & Teisman 2003). In addition, some parties may lack administrative capacities and/or financial resources to access or fully participate in partnerships. Derkzen and Bock (2007) mention that some parties may also lack the necessary social and symbolic capital to access and participate meaningfully in PPPs. In a study of rural development in the Netherlands, they noted how certain parties were labeled as professionals and seen as experts. Their input was considered more valuable, whereas representatives of local farmers' organizations saw their knowledge devalued. This does not only appear to apply to individuals: some organizations and institutions may also be considered more professional because they conform to dominant norms about how "proper" organizations and institutions are organized, regulated, and operated (Brunsson & Sahlin-Andersson 2000). Local actors will sometimes have certain organizational and institutional models forced upon them in order to be able to participate in PPPs, which can lead to conflicts with the constituencies of those actors who no longer trust the newly created organizations or institutions. In other cases, the formalization of organizations and institutions renders them accessible only to the elite (Mosse 2004). Nevertheless, power relations are dynamic, and local or civil-society organizations may increase their social and symbolic capital over time (Derkzen & Bock 2007).

Representation and accountability are other issues that need to be taken into account in relation to PPPs. When public institutions enter into a PPP, what happens to the public control over their activities and goals? It is likely that the distribution of the (beneficial) outcomes of PPPs reflect power relations: they are not neutral tools realizing win-win situations for all the partners involved.

A related issue is that it is not always clear which organizations and institutions are public and which are private (Starr 1988; Entwistle & Martin 2005). The privatizations of the 1980s and the participation of public institutions in PPPs have resulted in further blurring of the boundaries, sometimes deliberately, by private actors gaining control over public institutions, as Wedel

(2003) showed in her study of privatization and PPPs in postsocialist Russia. Starr (1988) argued that the way the public and private sectors and institutions were constituted varied from country to country, depending on its institutional history. The same applies to defining public and private goods, which is especially relevant in the case of agriculture and land reforms, as is described below. He therefore concluded that general statements about the effects of privatization and PPPs are hard to make. Yet other authors, such as Klijn & Teisman (2003), argue that because there is such a clear separation between the public and the private sector, especially in terms of what they refer to as organization culture, it is difficult to make PPPs work. Nevertheless, what does emerge from close reading of the literature on PPPs is that it is important not to take them at face value. No two PPPs are identical and careful study of power relations, goals, interests and modes of organization and operation, and scale of operation is needed, as is an analysis of the institutional context in which PPPs are operating in order to understand the distribution of benefits, costs, and risks among partners.

PPPs in Land Reform in South Africa: Strategic Partnerships in Limpopo Province

After the transition to democracy in 1994, the South African government adopted an ambitious and wide-ranging land-reform policy consisting of several subprograms: (i) land restitution, which allowed communities and individuals who had lost their land as a result of discriminatory legislation to reclaim it; (ii) land redistribution that assisted historically disadvantaged groups and individuals in obtaining land to foster a more equitable distribution; (iii) tenure reform aimed at securing land rights for those members of historically disadvantaged groups living on commercial farms, in the former homelands, or those who hold land in communal tenure. The land-reform policy was underpinned by constitutionally guaranteed rights to land restitution and land-tenure security. Over the years, however, the land-reform program stagnated and a shift took place toward greater dominance of the market and commercial farming. At first, this shift was especially pronounced in the redistribution component of land reform. Land needed for this component was always bought on a "willing seller, willing buyer" basis with prices set by the market, but what changed was the assistance provided by the state to land-reform beneficiaries (Lahiff 2003; Hall 2004). Initially, the Settlement/Land Acquisition Grant (SLAG) provided grants for poor people to help them access land for subsistence purposes, but land prices have always been high and grants were often insufficient to both obtain land and invest in agricultural production. With the adoption of the government's neo-liberal macroeconomic Growth Employment and Redistribution (GEAR) strategy, the basic principles of which continue to be important in the new (2006) Accelerated and Shared Growth Initiative South Africa (ASGISA) program, subsidies, protection, and other support to agriculture have been severely cut (Tilley 2002; Mayson 2003). As a result, land-reform beneficiaries face substantial obstacles in engaging in agricultural production. The Department of Agriculture and Land Affairs replaced SLAG with the Land Redistribution for Agricultural Development (LRAD) policy that made larger grants available, but mainly to those able to contribute to the investment in land and agricultural production. Hall *et al.* (2003: 5) argued that although LRAD supposedly contributes to the development of a range of agricultural developments from subsistence to commercial farming, in practice the program favors commercial farming by those with substantial assets. According to Mayson (2003), for those with fewer assets, the new approach renders partnerships attractive with the private sector in the form of joint ventures, and such partnerships are actively promoted by the South African government (Mayson 2003).

The shift toward emphasizing commercial farming is also notable in cases of restitution. In Mpumalanga and Limpopo Provinces, the government was struggling with restitution claims on farms producing high-value export crops, fearing that returning the land to claimant communities would result in a drop in production and export revenues. It is in this province in particular that the new model of strategic partnerships between land-reform beneficiaries and the private sector was embraced with enthusiasm (Derman *et al.* 2007; Fraser 2007). This new model stipulates that successful claimants, organized in a Communal Property Association (CPA) or trust, had to form a joint venture with a private-sector entrepreneur. The entrepreneur would invest working capital and take control of all farm management for a period of ten years, with the option of renewal for another period (*Ibid.*: 2–3). The idea behind the model is that it provides land-reform beneficiaries not only with capital to invest in agricultural production but also with the expertise of commercial farmers or private companies (*Ibid.*: Mayson 2003). The entrepreneur is supposed to train land-reform beneficiaries in how to operate a successful commercial farm and ensure that the beneficiaries receive a profitable and functioning farm when the contract ends. According to the Terms of Reference developed for the accreditation of strategic partnerships (DLA 2008: 5), experience in capacity building is one of the criteria used in selecting private-sector partners.

Many government officials and organizations representing commercial farmers show a deep distrust of smallholder farming, considering the commercial farm model far superior. In interviews[1] in Limpopo Province, the issue of land reforms and restitution generated fears concerning the "vandal-izing" and "destruction" of farms handed over for restitution or redistribution. A report published by the Centre for Development and Enterprise in South Africa stating that land reform leads to a decrease in production received a lot of publicity (*Mail & Guardian* May 6, 12 & 13, 2008). The report claims that 50% of all land-reform projects failed and related this to the priority that government had allegedly until recently accorded to small-scale produc-tion in land-reform projects. It cites the Director General of the Land Affairs Department who warned of "assets dying in the hands of the poor" (*Mail & Guardian,* May 6, 2008). Critics of the report argue, however, that while indeed there may be problems with the productivity of land-reform schemes, the assumption that government has favored small-scale production in land reform is not correct and that, on the contrary, small-scale producers on resti-tuted/redistributed land are hampered by "inappropriate large-scale models of agriculture foisted on to them by government officials and consultants. With the absence of post-settlement support, this is a key reason for the high failure rate in land reform" (Fraser 2007; *Mail & Guardian,* May 13, 2008). Detailed case studies of small-scale farming in Limpopo Province revealed that, despite these obstacles, most land-reform beneficiaries have seen improvements in their livelihood, though not as much as they expected when they joined the land-reform schemes (Lahiff *et al.* 2008).

Before land is handed over, the CPAs or trusts have to develop a busi-ness plan in cooperation with government agricultural extension officers or consultants. The business plans that were developed for each of the schemes studied were deemed unrealistic in terms of predicted economic returns and estimated costs (Lahiff *et al.* 2008). One of the problems experienced with business plan development is that the parties involved often lack clear infor-mation about the situation on the properties that will be handed over, such as the state of the farm and the availability of agricultural equipment (Bos 2009). Although the beneficiaries themselves were positive about improve-ments to their livelihoods, the Department of Agriculture is concerned about allegations that land reform is leading to decreased production, and has started a process of deregistering members of CPAs and trusts who are judged as not participating in production. Many active members fear that they may be judged as insufficiently productive, especially if their performance is eval-uated against the highly unrealistic business plans, and that they may be deregistered against their will (*Ibid.*: 62). Apart from the fact that there is lit-tle legal basis for the government to deregister members of CPAs and trusts, Lahiff *et al.* (2008) also argue that no drop in production has taken place since in the cases studied and the farms concerned had been left idle for an extensive period of time before they were handed over to the beneficiaries.

This has been the case with many farms that were offered to the government for restitution or redistribution by their owners. Nevertheless, the Director of the Centre for Development and Enterprise (CDE) is calling for a change in land reform: "We are proposing a public-private partnership to provide the leadership South Africa needs to show that we can resolve a difficult issue arising from our history and do it in such a way that everyone benefits from the process" (www.cde.org.za). The new model of strategic partnerships adopted in Limpopo Province preceded this call. It fits a long historical tradition of distrust in small-scale producers in South Africa, and many neighboring countries too (Spierenburg 2004; Hughes 2006).

The model of strategic partnerships is presented as *the* solution that will offer justice to the landless and contribute to poverty alleviation while still maintaining high levels of production. The assumptions underlying the model, however, should be questioned. The high levels of productivity before transfer are assumed rather than ascertained. Though further study is needed, several staff members at the organizations involved in assisting land-reform beneficiaries have complained that the trees in quite a number of the citrus groves offered for partnerships in land reform needed replacement.[2] This is consistent with earlier findings concerning the lack of quality and the productivity levels of the farms offered for land reform under the "willing seller, willing buyer principle" (Hall 2004: 18).

Another more implicit assumption is that strategic partnerships are "real" partnerships in which all the partners are equal and have mutual goals (cf. Brinkerhoff 2002). However, private partners and land-reform beneficiaries often harbor suspicions regarding each other's objectives which may influence the implementation of the strategic partnership (Bos 2009). Some of the strategic partners interviewed had serious doubts about the beneficiaries' interests in farming, arguing that they often seem not to have a vision concerning the use of the land. "They put the claim in for a piece of land, not for a business," remarked one of the partners interviewed by Bos (2009: 59). A study conducted by Ntsholo (2009) at the Moletele Land Claim in Limpopo shows the deep-seated mistrust of white farmers that is harbored by many of the beneficiaries as a result of years of discrimination and exploitation. One of the community respondents expressed his concerns as follows:

> We could not go to school because of the white farmers, our parents died because of the white farmers, we as the elders had to lead senseless lives because of these white farmers. We are suffering, our children are suffering and I attribute all that to the white farmers. They cannot be trusted, they are evil people. So going into business with them is a very bad idea.
>
> (cited in Ntsholo 2009: 121)

No attention is paid to power relations between the private sector or commercial farmers on the one hand and the land-reform beneficiaries on the other, or within these groups. Ntsholo (2009) noted that the relationship between private-sector partners and the beneficiaries of land reform was often perceived as condescending by beneficiaries, which was attributed to years of "master-servant" social constructs between the white farmers and their black employees. The role of government officials in mediating between the groups appears to be limited, and many seem to be biased toward the commercial farm model. Furthermore, one of the most consistent complaints about land reforms concerns the general lack of government capacity and funding to assist land-reform beneficiaries (Hall *et al.* 2003; Lahiff 2003; Hall 2004). Coordination between the Department of Land Affairs, which arranges the transfer of land, and the Department of Agriculture, which has to approve the business plans, is frequently problematic. The strategic partners interviewed by Bos (2009) indicated that increasingly the Department of Agriculture is driving the development of strategic partnerships and this is having a positive impact on the partnerships. Nevertheless, all six private partners and the CPA members that Bos interviewed were very clear that the lack of financial support, and especially the failure to deliver grants on time, was a significant problem that was hampering the success of the partnerships. The delays in funding often result in a shift in focus from development and training to maintaining a bare minimum level of operation on the farms. This in turn, as the private partners themselves indicate, feeds distrust on the part of land-reform beneficiaries regarding the motives of the private-sector partners (Bos 2009).

The potential benefits of strategic partnerships for beneficiaries, as cited by Derman *et al.* (2007), include rent for use of the land paid by the private-sector partner, a share of the profits, preferential employment, training opportunities, and the promise that they will receive profitable and functioning farms at the end of the contract/lease agreement. However, it is questionable whether the beneficiaries—or rather their CPAs or trusts—will be able to ensure that training is part of the business plan or that they have the capacity and means to put leverage on the private-sector partner to honor promises of training. Bos's recent survey (2009) showed that only two of the six strategic partnerships she studied had training programs in place, and in only one case did these extend beyond direct production capacities. Beneficiaries often need training not only on the production side but also in management and marketing products, which is an area that is often neglected. These partnerships may result in job opportunities but it may well be that old relations of production will continue with only low-paid jobs available for a segment of the community of beneficiaries. Furthermore, as some have warned, if the laborers belong to the beneficiary community, they are share-holders, and private-sector partners may therefore argue that labor legislation pertaining to working conditions and the minimum wage does not apply.[3] There is also the issue of what happens to farm laborers who were working on the farm before the transfer but who are not part of the beneficiary community. Derman *et al.* (2007) found at least one case in which all former laborers had been fired when the farm was handed over to the claimant community. Further study is needed to gain a better understanding of the labor issues in strategic partnerships.

Commercial farmers and companies have an interest in developing joint ventures, given the cutbacks in government support for agriculture that are affecting commercial farmers as well. Mayson (2003) cites a number of reasons for entering into strategic partnerships: Firstly, there is a need to restructure farmers' and companies' operations. It appears that many commercial farmers in Limpopo Province are withdrawing from the production side and are moving into marketing. Five of the six strategic partners participating in the study by Bos (2009) owned subsidiary organizations in agriculture—both upstream and downstream in the commodity chain—without owning land. By engaging in a partnership, the potentially most risky part of the chain, that is, production, is allocated to the beneficiaries. If training and participation in marketing is not granted to the beneficiaries, they will bear the highest risk in the process (Derman *et al.* 2007), though the distribution of risks along the value chain may vary from crop to crop. The private-sector partner often obtains a management fee (which is more or less guaranteed as long as turnover can be maintained), a share of the company profits, and exclusive control of upstream and downstream activities, with potential benefits exceeding those of the farming enterprise itself. Furthermore, by entering into partnerships with multiple communities in a specific area, with each owning numerous farms, the private-sector partners have the chance to consolidate and rationalize production in a way that was not generally open to the previous owner-occupiers (*Ibid.*: 12). While the strategic partner is required to share profits from on-farm production with the communities, no such requirement applies to other parts of the value chain over which the strategic partner has exclusive control (*Ibid.*: 14). Critics of the partnership model therefore warn that joint ventures are mainly ways for white commercial farmers and companies to spread the risk of engaging in an increasingly complex and capital-intensive sector, while at the same time gaining political credibility (Mayson 2003). Just how risky a strategic partnership may be became clear when one of the most prominent private-sector partners in Limpopo Province, which was involved in a number of strategic partnerships in the province, went bankrupt in 2009. While a forensic audit is still being conducted, rumors, which were strenuously denied by the company's spokesperson, circulated that the company had concocted the bankruptcy by shifting money from the partnership to its various subsidiary companies (*Farmers Weekly* April 2009). Hellum & Derman (2009) argue that the claimants had become dependent on the partner, who charged the partnership high fees for farming inputs and the use of equipment. The company had managed to acquire a loan of SAR[4] 5 m on behalf of the community of beneficiaries, and now that the company is officially bankrupt, the community cannot repay this debt and has no funding left to invest in the farms (*Sunday Times,* March 1, 2009).

Land reform also offers opportunities for accessing capital support from the state for the expansion of production and CSR, including development funds, grants, and other support provided to land-reform beneficiaries. For example, beneficiaries receive a sum of between SAR 1,500 and SAR 3,000 per household for the development of the land. Given that the model of strategic partnerships applies to large-scale farms, the claimed areas pertain to hundreds, in some cases thousands of beneficiary households, though the government has been slow in paying these grants.

This last reason may, in particular, lead to imbalances in the partnership. In most joint ventures the private-sector partner receives 48% of the shares, the Communal Property Association of Community Trust of the beneficiaries gets 50%, and farm workers who are not part of the beneficiary community get 2%. However, CPAs or trusts can apply for government grants to support the development of their enterprise. It may therefore very well be the case that their contribution to the assets in the form of land as well as grants exceeds 50%, and that the private-sector partner contributes far less than their percentage of the shares justifies. Yet this division of the shares has become standard practice and is applied without a detailed review of what each of the partners contributes (Derman *et al.* 2007).

The fact that the whole process of transferring the land to the beneficiaries and the approval of the partnerships takes so long, sometimes up to three years, is proving to be a major obstacle. Former owners who are unsure of the outcome of the process will not invest in farm maintenance, which results in huge investment needs once the farm is transferred. Grants to beneficiaries take even longer to be transferred, rendering these investments difficult. In some cases, the partners are the former owners, who may be short of funding to make the investments needed. Commercial banks appear not to know how to deal with restituted/redistributed farms and strategic partnerships, and some are reluctant to offer loans if it is not clear whether the land can serve as collateral, perhaps even more so following the above-mentioned bankruptcy case.[5] Sometimes beneficiaries run the risk of a debt trap: the private-sector partner offers an advance payment to be reimbursed once the grant is paid but at a high interest rate. Beneficiaries may feel forced to accept this because otherwise assets on the farm will be neglected while the partners are waiting for the grants, making it more difficult to restart the farm once the grants arrive.

Negotiating contracts is difficult and many CPAs and trusts do not have the capacity to do so, while many private-sector partners have extensive legal and financial experience and may engage the services of well-trained lawyers. Nkuzi, a land-rights NGO in Limpopo Province, has used the services of a well-reputed private law firm to assist beneficiaries on a pro bono basis but the firm is based in Johannesburg, the time available is limited, and the problems that the beneficiaries are experiencing are numerous. Furthermore, the beneficiaries are not always used to the fact that they have to actively approach lawyers, and that they, as clients, need to instruct them.[6] A case involving the Makuleke claim illustrates the potential difficulties. The Makuleke have successfully claimed an area in the Kruger National Park (in the Limpopo Province part of the park) from which the community was evicted in 1969. The Makuleke CPA manages its part of the Kruger Park through a Joint Management Board on which representatives of the South African National Parks (SANParks) and the Kruger National Park management also sit. The CPA has been granted the right to commercially exploit their part of the Kruger, though their activities are subject to environmental impact assessments (Spierenburg et al. 2006, 2008). One of the first steps taken by the Makuleke was to establish a highly profitable hunting camp on their land, which they used for a limited number of high-profile hunts per year. As a second step, an agreement was made with a private-sector partner to develop a game lodge, called The Outpost, on the western section of their land. Recently, however, the Makuleke signed a surprisingly unfavorable agreement with another safari operator, Wilderness Safaris. This concession will last for 45 years, which is a long time, especially considering that the contract does little to hold the private-sector partner to a certain level of performance and does not contain clear exit clauses that would allow the Makuleke to extract themselves from an unprofitable relationship. It also effectively prevents the Makuleke from hunting on the land. The community did have access to competent legal advisors. From 1997 onward, an NGO-like group called the Friends of Makuleke (FoM) provided the community with technical expertise in the land-claims process, supporting the community's Legal Resources Centre attorney and, as such, played an important role in the success of the claim. The FoM was disbanded shortly before the signing of the contract but some of its former members have continued to advise the Makuleke. Responses from former members were mixed. One felt that this was the best deal that the Makuleke were likely to get, while another advised the Makuleke not to sign. However, this negative advice came one day before the signing ceremony and was not followed up. The game lodge currently generates less than what was generated by the hunting operation and it remains to be seen whether Wilderness Safari's higher projected income will be achieved (Spierenburg *et al.* 2006).

A final potential problem with the strategic partnership model is that of power relations within the beneficiary community and differences in visions about the use of the land resulting from socioeconomic differentiation. Research by Hellum & Derman (2009) shows that the CPAs or trusts, though these should be democratically elected, do not always represent the beneficiary communities (Lahiff *et al.* 2008; Ntsholo 2009). Elites may dominate the negotiations about the partnerships, and they have different interests than the poorer members of the communities. Power disparities within beneficiary communities often result in elites presenting themselves as the legitimate voice of the poor, while the whole private-sector community arrangement is supposed to benefit the poor members of the communities but their voice is not included (Ntsholo 2009). For instance, when the chairman of the CPA involved in the Hoedspruit Claim, who has a salaried job, was asked whether all the members of the community were in favor of the partnership or whether some might prefer to move on to the land themselves to farm, he replied: "Yes, unfortunately we have many people who want to farm. In the homeland we had irrigation schemes, I think that is why, but these schemes were abandoned by government years ago." One of the partners who owned part of the claimed land added: "People have large tracts they do not use. We should not move people here, but we should get the infrastructure back in so they can develop where they are now and use the funds generated by the company to do that. The irrigation schemes were left in ruins, but they can be rebuilt. We sometimes forget the potential in those areas."

This problem with representation also points to another problem with PPPs, namely, the difficulties in establishing what is public and what is private (Starr 1988; Wedel 2003). In theory, CPAs and trusts are democratically elected local government bodies. Hence, a partnership between a CPA or trust and a private-sector company can be rightfully termed a PPP. In some cases, however, the CPA or trust appears dominated by a few individuals who have a personal interest in the partnership, and this partnership may then closely resemble a PPP, though government is still involved in approving the land transfer and the partnership. In Bos's (2009) study, various strategic partners admitted that the CPAs did not always represent the community and, consequently, that other community members do not always benefit from the PPP.

This may lead to new forms of exploitation, especially given the danger that labor legislation is considered by the partners not to apply to "shareholders."

Dealing with power relations within strategic partnerships is by no means easy, especially since in some cases there are also internal power relations at play amongst the group of land-claim beneficiaries. The strategic partnership model is one in which complex legal and business matters are at stake. The private-sector partners, mostly experienced corporate players with extensive legal and financial experience, are in a stronger position to influence the terms of the contracts than the Land Claims Commission or the land-reform benefi-ciaries (Spierenburg *et al.* 2006; Derman *et al.* 2007: 10). As described above, certain land-rights NGOs are assisting claimant communities by engaging the services of lawyers and paralegal assistants. Derman *et al.* (2007: 10) describe how, in one partnership, lawyers proposed numerous changes to the contract that did not change the model but were an attempt to ensure that communi-ties have greater control over the joint operations as well as access to unused portions of the farm. However, the contract has not yet been signed. Access to lawyers remains difficult and for many beneficiaries it is not easy to direct lawyers or check the quality of their work (Spierenburg *et al.* 2006, 2008).

Derman *et al.* (2007) do attribute threats not only to the better skills and more extensive experiences of the private sector but also to the speed with which the model was being implemented, particularly in Limpopo Province. The Regional Land Claims Commission and the Provincial Department of Agriculture developed the model without consulting the claimant commu-nities. Bos (2009) indicates that expectations were not set at the start of the partnership, which might lead to disappointment and mistrust between stake-holders during the period of cooperation. The need for a feasibility study at the start of the partnership was often mentioned by the strategic partners. One form of protection that is provided by the state is the clause in the restitution contract stating that beneficiaries may not sell their land for a period of ten years. But whether this clause can in fact protect communities from building up debts with their strategic partners that could force them to sell the land after the clause expires is questioned.

Conclusions

This chapter has described the emergence of the PPP concept as a panacea for improving service delivery and economic development. By forging stronger connections between the state and the private sector, and allowing more market influence in the public sector, service delivery to citizens is sup-posed to improve and economic development should be stimulated. Through international financial institutions, this approach has become prominent in the South (Batley 2004), bringing it into the dominant neo-liberal world order. South Africa has enthusiastically embraced the neo-liberal approach (Terreblanche 2002), which appears to have indeed resulted in high levels of economic growth but at the same time in a widening gap between the rich and poor.[7]

The PPP model has been adopted as a solution to perceived problems with land reforms in South Africa, especially in Limpopo Province. The idea was that connecting land-reform beneficiaries and private-sector partners could serve several purposes at once: It could contribute to social justice by enhancing the (land) rights of the formerly disadvantaged, to the transformation of South Africa's rural economy by changing the racially skewed distribution of land, and to the maintenance of agricultural production. However, as this contribution has aimed to demonstrate, there are a number of contradictions. First of all, questions have been raised about the extent to which the strategic partnership model is capable of transforming the rural economy. The risks associated with the model are that large-scale farms will continue with "business as usual" with most of the benefits accruing to the private-sector partners and few to land-reform beneficiaries, and labor relations will remain basically the same. Second, the connections between the private sector and communities of land-reform beneficiaries could include certain sectors of these communities, but may also exclude others. As this chapter has shown, the local government institutions involved in the partnerships do not always represent the whole community, and seem to serve the interests of an elite that is not interested in returning to the land. Information about the functioning of the partnership does not seem to reach far beyond the group represented in these institutions.

What is striking is the confidence that governments appear to have in the efficiency and effectiveness of the private sector. The strategic partners are supposed to provide training to land-reform beneficiaries to prepare them to take over the farms but it is unclear if and how they will be held accountable for accomplishing the training. There appears to be no critical examination of the notion that the commercial farming model is the only viable land-use option. This model is, as Fraser (2007) also remarks, being forced upon land-reform beneficiaries in Limpopo Province. Belief in the private sector appears to be so strong that one of the lessons learnt from more "conventional" development programs, namely, the need for feasibility studies, monitoring, and evaluation, appears to have been forgotten. This is all the more remarkable as the private sector is believed to be much more outcome driven than the public sector (Brunson & Sahlin-Anderson 2000).

Another lesson learnt from conventional development programs concerns community participation. Many studies have shown that community participation in development projects is no easy feat but is still a necessary and integral part of the implementation phase of a development project and, more importantly, in the project design phase too (Agrawal & Gibson 1999; Ribot 1999). Yet in the land-reform program in Limpopo Province, the needs of beneficiaries are not being taken into account. The strategic partnership model was a reaction to the earlier failure of some of the land-restitution projects that were attributed to a lack of support for land-reform beneficiaries (Hall & Lahiff 2003). However, the strategic partnership may not be the only—or for that matter the best—solution to dealing with land-reform failure. No attempts have been made to investigate the possibilities for supporting small-scale farming. In theory, land-reform beneficiaries are involved in the development of the business plans for farms. These beneficiaries are not, however, a homogeneous group and there are indications that the partnerships are being "captured" by the elite members of the beneficiary groups, which is not an uncommon problem in development projects (Mosse 2004; Platteau 2004). The connections with private-sector partners are necessarily limited, given the large number of beneficiaries involved in each of the claims. And it appears that only a few elite members of the beneficiary community are in fact benefiting from these connections. If participation is to be meaningful, local farmers and land-reform beneficiaries need to be empowered in terms of capacities and skills to negotiate with private-sector partners and develop business plans. As the example of the Makuleke claimants and their negotiations with tour operators shows, (contract) negotiations can be extremely complex and many private-sector companies have an advantage over local farmers and beneficiaries in terms of capacities and access to legal services. In addition, benefits for local farmers and land-reform beneficiaries also depend on their position in the value chain and the distribution of risks and profits along that chain. Providing local farmers and beneficiaries with insights into their relative position in the value chain is crucial.

Finally, this case confirms that it is not always clear in PPPs what the public and private interests and/or roles of the different parties involved are. CPAs entering into partnerships with the private sector are, in theory, democratically elected local governance institutions and, hence, public institutions. Yet there is a risk that the members of the CPAs, who often belong to the elite and are advantaged vis-à-vis other community members in terms of education, will defend the interests of the elite rather than other members of the community of land-reform beneficiaries and will begin to act as if they are forming a private company. Participation by a public institution in a PPP may also shift the interests of the institution toward the success of the PPP rather than taking the public interest into consideration. As Klijn & Teisman (2003) concluded from their analysis of PPPs in the Netherlands, it should be clear before embarking upon a PPP what the interests of the different parties involved are and what their roles are, not only within the PPP but also regarding their mandate. It is crucial that this mapping of interests and roles continues at regular intervals as they may change over time. PPPs can too easily be considered win-win strategies, and differences in approach, mandate, and interests are, as a result, not always communicated transparently between and to all the parties involved.

Notes

1. These interviews were conducted in the context of a joint research project in the Programme for Land and Agrarian Studies in South Africa and the VU University of Amsterdam in which one of the authors is involved. This project is funded by the South African Partnership for Alternative Development (SANPAD) and the VCAS Association of the VU University in cooperation with the VU Centre for International Cooperation.
2. Interviews with staff members of the Rural Action Committee, Nelspruit, November 2007, and MABEDI, Bushbucksridge, November 2007.
3. Interviews with staff members of the Rural Action Committee, Nelspruit, November 2007.
4. SAR = South African Rand. In 2009 EUR 1 fluctuated between SAR 11 and SAR 12. See also the website of the Department of Land Affairs, Government of South Africa: www.gov.za
5. Interviews with members of the CPA and partners involved in the Hoedspruit claim, November 2007; interviews with staff members of the Rural Action Committee, Nelspruit, November 2007.
6. Interviews with staff members of Nkuzi, Makhado, February 2008.
7. South African Institute of Race Relations, "South African Survey Online—Business and Employment," accessible through: http://www.sairr.org.za/services/publications/south-africa-survey/south-africa-survey-online-2007-2008/2007-2008-business-and-employment/?searchterm=Business%20and%20 Employment. See for a discussion of this trend, Seekings & Nattrass (2005).

References

Agrawal, A. & C. C. Gibson (1999) "Enchantment and Disenchantment: The Role of Community in Natural Resource Conservation," *World Development* 27(4): 629–649.

Atkinson, D. (2007) *Going for Broke: The Fate of Farmworkers in Arid South Africa.* Cape Town: HSRC Press.

Batley, R. (2004) "The Politics of Service Delivery Reform," *Development and Change* 35(1): 31–56.

Bos, A. (2009) "Government, Land Reform Beneficiaries and Private Organizations; Join Hands in the Struggle for Land Restitution? Experiences of Strategic Partners and Third Parties Involved in Public-Private Partnerships within the Restitution Programme in South Africa," Unpublished M.Sc. Thesis, VU University Amsterdam.

Brinkerhoff, J. M. (2002) "Government Non-profit Partnership: A Defining Framework," *Public Administration and Development* (22): 19–30.

Brunsson, N. & K. Sahlin-Andersson (2000) "Constructing Organizations: The Example of Public Sector Reform," *Organization Studies* 21(4): 721–746.

CDE (Centre for Development and Enterprise) (2005) "Land Reform in South Africa. A 21st Century Perspective," *CDE Research*, Policy in the Making Series 14. Johannesburg: CDE.

Derkzen, P. & B. B. Bock (2007) "Identity: Symbolic Power in Rural Partnerships in the Netherlands," *Sociologia Ruralis* 47(3): 189–204.

Derman, B., E. Lahiff & E. Sjaastad (2007) "Strategic Questions about Strategic Partners: Challenges and Pitfalls in South Africa's New Model of Land Restitution," Paper prepared for the Programme for Land and Agrarian Studies, University of the Western Cape.

DLA (2008) "Terms of Reference for the Accreditation of Strategic Partners for the Land Reform Programme," Tshwane: National Department of Land Affairs.

Edwards, M. & D. Hulme (1996) "Too Close for Comfort? The Impact of Official Aid on Nongovernmental Organizations," *World Development* 24(6): 961–973.

Entwistle, T. & S. Martin (2005) "From Competition to Collaboration in Public Service Delivery: A New Agenda for Research," *Public Administration* 83(1), 233–242.

Ferguson, J. (1990) *The Anti-Politics Machine.* Cambridge: Cambridge University Press.

Fraser, A. (2006) "Geographies of Land Restitution in Northern Limpopo: Place, Territory, and Class," Ph.D. Thesis, Ohio State University. http://etd.ohiolink.edu/send-pdf.cgi?osu1148498881, accessed June 17, 2009.

Fraser, A. (2007) "Hybridity Emergent; Geo-history, Learning and Land Restitution in South Africa," *Geoforum* 38: 299–311.

Hall, R. (2004) "A Political Economy of Land Reform in South Africa," *Review of African Political Economy* (100): 213–227.

Hall, R., P. Jacobs & E. Lahiff (2003) "Evaluating Land and Agrarian Reform in South Africa: No. 10," Final Report. Cape Town: Programme for Land and Agrarian Studies, University of the Western Cape.

Hellum, A. & B. Derman (2009) "Government, Business and Chiefs: Ambiguities of Social Justice through Land Restitution in South Africa," in: F. von Benda Beckmann, K. von Benda Beckmann & J. Eckert (eds.), *Rules of Law and Law of Ruling. On the Governance of Law.* Farnham: Ashgate, pp. 125–150.

Hughes, D. M. (2006) *From Enslavement to Environmentalism: Politics on a Southern African Frontie.* Seattle: University of Washington Press.

Klijn, E-H. & G. R. Teisman (2003) "Institutional and Strategic Barriers to Public-Private Partnership: An Analysis of Dutch Cases," *Public Money & Management* (July): 137–146.

Lahiff, E. (2003) "Land and Livelihoods: The Politics of Land Reform in Southern Africa," *IDS Bulletin* 34(3): 54–63.

Lahiff, E. (2008) "Land Reform in South Africa: A Status Report 2008," *Programme for Land and Agrarian Studies*, Cape Town.

Lahiff, E., T. Maluleke, T. Manenzhe & M. Wegeriff (2008) "Land Redistribution and Poverty Reduction in South Africa: The Livelihood Impacts of Small-holder Agriculture under Land Reform," Research Report 36, Cape Town: Programme for Land and Agrarian Studies.

Linder, S. H. (1999) "Coming to Terms with the Public-Private Partnership: A Grammar of Multiple Meanings," *American Behavioral Scientist* 43(1): 35–51.

Mayson, D. (2003) "Joint Ventures," Cape Town: Programme for Land and Agrarian Studies, University of the Western Cape.

Monaci, M. & M. Caselli (2005) "Blurred Discourses: How Market Isomorphism Constrains and Enables Collective Action in Civil Society," *Global Networks* 5(1): 49–69.

Mosse, D. (2004) "Is Good Policy Unimplementable? Reflections on the Ethnography of Aid Policy and Practice," *Development and Change* 35(4): 639–671.

Ntsholo, L. (2009) "Land Dispossession and Options for Restitution and Development: A Case Study of the Moletele Land Claim in Hoedspruit, Limpopo Province," Unpublished M.Phil. Thesis, University of the Western Cape.

Pauw, K. (2005) "A Profile of the Limpopo Province: Demographics, Poverty, Inequality and Unemployment," Provide Project Background Paper 1 (9). Elsenburg: Provide Project.

Platteau, J-P. (2004) "Monitoring Elite Capture in Community-driven Development," *Development and Change* 35(2): 223–246.

Ramutsindela, M. (2003) "Land Reform in South Africa's National Parks: A Catalyst for the Human-Nature Nexus," *Land Use Policy* 20: 41–49.

Ramutsindela, M. (2007) "The Geographical Imprint of Land Restitution with Reference to Limpopo Province, South Africa," *Tijdschrift voor economische en sociale geografie*, 98(4): 455–467.

Roe, E. (1991) "Development Narratives, or Making the Best of Blueprint Development," *World Development* (19): 229–232.

Roe, E. (1995) "Except-Africa: Postscript to a Special Section on Development Narratives," *World Development* 23(6): 1065–1069.

Seekings, J. & N. Nattrass (2005) *Class, Race and Inequality in South Africa*. New Haven & London: Yale University Press.

Spierenburg, M. (2004) *Strangers, Spirits, and Land Reforms: Conflicts about Land in Dande, Northern Zimbabwe*. Leiden: Brill.

Spierenburg, M., C. Steenkamp & H. Wels (2006) "Resistance against the Marginalization of Communities in the Great Limpopo Transfrontier Conservation Area," *Focaal, European Journal of Anthropology* (47): 18–31.

Spierenburg, M., C. Steenkamp & H. Wels (2008) "Enclosing the Local for the Global Commons: Community Land Rights in the Great Limpopo Transfrontier Conservation Area," *Conservation and Society* 6(1): 87–97.

Starr, P. (1988) "The Meaning of Privatization," *Yale Law and Policy Review* (6): 6–41.

Stephenson, M. O. Jr. (1991) "Whither the Public-private Partnership: A Critical Overview," *Urban Affairs Quarterly* (27): 109–127.

Terreblanche, S. (2002) A *History of Inequality in South Africa, 1652–2000*. Pietermaritzburg: University of Natal Press.

Wedel, J. R. (2003) "Clans, Cliques and Captured States: Rethinking Transition in Eastern Europe and the Former Soviet Union," *Journal of International Development* 15: 427–440.

Newspaper articles

Farmers Weekly, March 27, 2009, "South African Farm Management's Side of the Story."

Mail & Guardian, May 6, 2008, "Warning on Deteriorating Land Restitution Process."

Mail & Guardian, May 12, 2008, "Why Land Reform is Stuck."

Mail & Guardian, May 13, 2008, "Food vs. Land Reform."

CHAPTER 9

Connectivities Compared: Transnational Islamic NGOs in Chad and Senegal

Mayke Kaag

Introduction: Connecting to the *Umma* through Islamic Relief

A remarkably persistent idea is that Africa only became connected to the outside world with the arrival of European traders in the sixteenth century and during the subsequent colonial conquest. This kind of Eurocentrism prevents us from recognizing that Africa has in fact been connected to the Arab world for a long time, for instance, through Islamic scholars, trade, and people performing the *hajj*. Economic links, such as trans-Saharan trade, became weaker after colonial conquest but religious exchanges continued throughout the colonial era. The last 20 years have seen a reintensification of links between Africa and the wider Islamic world due to better transport and communication possibilities, the oil boom in the Gulf, and the efforts of postcolonial African governments to reinforce relationships beyond those ties inherited from colonialism (Hunwick 1997; Bennafla 2000).

Transnational Islamic relief organizations are part of this process of reintensifying African-Arab ties (Kaag 2008). However, the bridging by transnational Islamic charities is quite new in that it involves a specific understanding of Islamic solidarity, which is expressed through material aid and

This contribution summarizes earlier work published in *Afriche e Orienti* (Kaag 2010) and the *International Development Planning Review* (Kaag 2011b) and the author would like to thank the editors of both journals for granting their permission to use part of the articles for the present contribution.

moral support fueled by the modern (mass) media and directly includes not only the elite but also poor groups.

This contribution explores in detail how connections are being established through the work of these organizations. I will look first at the concept of Islamic solidarity, which is exemplified by *zakat* (obligatory almsgiving). I will show that the practice of *zakat* has increasingly been taking on a transnational connotation, which has contributed to the emergence of the transnational Islamic NGOs that are the subject of this chapter. I will then demonstrate how, through the charity chain of these NGOs, moral networks are being weaved. Connecting is, however, not only determined by these organizations' approaches and strategies but to a large extent also by the connectivity of the local contexts, that is, how the characteristics of the local contexts, including the strategies of local actors, do or do not facilitate the building and maintaining of connections with these NGOs. I will illustrate this argument by considering the reception of transnational Islamic NGOs in Chad and Senegal and comparing them regarding their connectivity and results in terms of the connections established.

As the term "connection" is so central to my argument, it is important to indicate at the outset what it may have to offer analytically. In my view, the advantage of using the concept of connection is that it draws attention to how links are established (connecting), how characteristics of actors and contexts on both sides influence whether and how links are established (connectivity), and points of exchange between different actors (connection). A connection presupposes that something is flowing: it is a concept that accentuates exchange, transmission, and flow, much more than concepts like relationship or link. In the following, I discuss how this will help us to capture processes involved in the work of transnational Islamic NGOs from the Arab world in Africa.

Islamic Solidarity in a Globalizing World

Central in the work of transnational Islamic NGOs is the idea of Islamic solidarity. It not only informs their activities but is also the basis of their funding. A large part of the funding of the majority of these organizations is provided by *zakat* or compulsory almsgiving.[1] *Zakat*, a form of financial worship, constitutes an act of social solidarity and an affirmation of faith (Benthall & Bellion-Jourdan 2003: 26). Other forms of pious gifting exist, such as *hadiya* (Soares 2005), in which the act of giving is of central importance and confers blessing (*baraka*) on the one who gives. What becomes of the money afterward is less important. In the case of *zakat*, however, the objective and final destination of the gift are also important: the wealthy Muslim gives to improve the lives of the poor. *Zakat* can thus be seen as a means of redistributing wealth in Muslim society (Weiss 2002). Furthermore, by linking pious givers to needy recipients, *zakat* can be considered a connection, a bridging device that enables flows of money and *baraka*, flows through which relationships are established.

The Qur'an states who is entitled to receive *zakat*. Traditionally, it used to be paid to the needy in one's own community but it has increasingly taken on a transnational connotation. This is, first of all, because communities have become more transnational as a result of international migration. In the case of the Senegalese Murid community, for instance, *zakat* used to be paid to the poor in one's own village or neighborhood, often through the local Imam. Nowadays, Senegalese Murid migrants from all over the world are sending *zakat* to their families back home to be used or distributed by them[2] (Kaag 2011c).

Second, it is being understood that the community can encompass the whole of the *Umma,* the global community of the faithful. This idea took root in wealthy countries such as Saudi Arabia and the United Kingdom when Muslims started to feel that the real poor no longer or only sometimes lived in their own countries, a feeling that has been strengthened by the mass media with their exposure of poverty and disasters worldwide (Benthall & Bellion-Jourdan 2003). It is this recognition that has stimulated the foundation of many of today's transnational Islamic NGOs, such as AMA (*Association des Musulmans d'Afrique*) and Islamic Relief.

A point of contention is whether *zakat* is for needy Muslims only or can also be given to non-Muslims. In general, it is considered to be for Muslims alone but if the concept is widened to include not only Muslims but also those who can become Muslim, then using *zakat* for aid to non-Muslims becomes appropriate (Benthall & Bellion-Jourdan 2003). For many Islamic NGOs from the Gulf region, this means that missionary activities or *da'wa* (the invitation or call to Islam) are and should be an important component in their strategy.[3]

Building Moral Networks through the Charity Chain

Islamic NGOs are Islamic in the sense that Islam is an important source of inspiration for them as organizations. Islamic NGOs may have differing objectives and areas of intervention but all share a foundation in the sacred textual sources of Islam, the Qur'an, and the Sunna (the authoritative practice of the Prophet Mohammed), and in the basic principles of Islamic law and ethics, which act on their identity and agenda, and how they obtain and distribute their resources. Due to the primordial role of faith in these organizations and the way they provide inspiration, guidance, and discipline, these NGOs can rightly be termed faith-based organizations (Bornstein 2002, 2003).

Islamic NGOs from the Arab world disseminate what is often called a Salafi form of Islam. Salafism is a modernist current that purports to follow the "pious predecessors" (Arabic: *salaf*), the first generation of Muslims whose practice of Islam it considers to be the purest form (Ghandour 2002). Salafis seek an Islamic revival through the elimination of what they consider foreign innovations (*bid'a*). Through their efforts to "reeducate" African Muslims about Islam and to purify Islam of allegedly un-Islamic practices (Rosander 1997), these NGOs are helping to ease processes of Islamization and Arabization, with the latter understood as an increased cultural orientation toward the Arab world that is expressed in the adoption of elements (such as language, styles of dress, and social and cultural values) of transnational Arab elite culture as a reference. Arabization and the teaching of Islam and Arabic are seen as an antidote to the effects of Western colonialism and contemporary influences from the West (Hunwick 1997). In these ways, connections to and within the *Umma* are thus built and reinforced.

Connections at the practical level are being built through the educational programs taught in schools that are financed and managed by some of these organizations. The curriculum is the official curriculum of the country of intervention, and lessons are given both in Arabic and in the official language of the country concerned. Wherever possible, however, an explicit Islamic perspective is adopted. History, for example, is taught from an Islamic viewpoint. This means that students in their knowledge and worldview become connected to the Arab world and its history. The transnational solidarity underlying projects is represented visually by the signboards placed outside the mosque or hospital stating the name of the donor from the Gulf or of the transnational charity both in Arabic and in the country's official language. Foster parents' programs for orphans link foster children and their families to sponsors in the Arab world and while the latter tend to remain rather abstract and anonymous to the target group, there is a sense that someone in the Arab world is involved and paying for them. This idea is also cultivated by the local supervisors of projects when Arab representatives come to the orphanages to distribute clothes and other materials. In the case of *da'wa*, the connection to the *Umma* is made still more explicitly: "You can become like us" was the message of a Saudi director of an Islamic charity during a visit to a village in southern Chad. While a racist undertone in this message might be noted,[4] what he intended to convey was that by converting to Islam, one becomes part of the *Umma* and all it stands for.

While the approach or social technology of these Salafi NGOs is basically the same all across Africa, that is, their instruments for bringing about their connection, their reception and thus the content of the connection may differ from one place to another depending on local dynamics and the issues at stake, and the way local people use these new modes of connectivity in their own strategies. This brings us to the importance of the local context.

Connectivity: The Importance of Local Contexts

There is a tendency in the West to perceive engagement and interventions in Africa by non-Western powers, such as China and the Arab world, from merely a geopolitical perspective, implying that these actors are only trying to enhance their position of power without considering what is good for Africa. This is the received idea in the West and characterizes its own engagement in the continent. Arab relief organizations do not escape this stereotyping and are often portrayed as the puppets of a Saudi master plan or as al-Qaeda's henchmen trying to make headway in Africa.[5] In more religiously orientated analyses, received ideas about African Islam (Sufi Islam, which is considered peaceful and tolerant) versus Islam in Africa (reformist Islam that is seen as a more violent externally introduced form of Islam) play a role (cf. Westerlund & Rosander 1997). Such geopolitical perceptions have the collateral of reducing Africa and Africans to the role of passive victims. In contrast to this view, this chapter argues that, while these Islamic relief organizations of Salafi background apply a similar approach wherever they intervene, the effects are different in different local contexts as a result of encounters between people, visions of Islam, and an intersection of strategies, which clearly shows that African target groups and other stakeholders cannot be considered passive recipients or even victims but are evidently coshaping transnational Islam. Building connections should be seen as a two-way process rather than a one-way flow.

This process of building connections is explored here by analyzing the interaction between transnational Islamic NGOs and their target groups in Chad and Senegal.[6] Comparing these two countries is interesting as they differ in important ways: in the role Islam is playing in the public sphere and in local and national power dynamics; in the characteristics of poverty; and in how "development" is organized locally, in particular the characteristics of civil society. How these factors are shaping the connectivity between the target groups and NGOs is explored below, as are the ways in which the interaction between the different actors is contributing to building new connections between African Muslims and the wider *Umma*, while also not ignoring processes of (local) disconnect.

Transnational Islamic NGOs in Chad

In June 2004, I visited an orphanage in N'Djamena that was being run by the Saudi relief organization al-Makka al-Mukarrama. It was a festive day as a Saudi representative had come to distribute clothes to the children who had been organized in rows and were clearly impressed by their important visitor from the Gulf. One by one they were called forward to receive their package of clothes that, they were told, had been donated by pious Muslims in Saudi Arabia. They should work hard and study well to become good and pious Muslims themselves. The Saudi representative approached the children and gave them their clothes with fatherly warmth while the whole event was filmed so it could be shown to donors back in Saudi Arabia.

This short vignette illustrates how connections are being weaved through the work of this Islamic NGO. At one level, a connection is being established by the flow of material aid to target groups and the return of a blessing to the donors, while at another level, images and moral messages are being transmitted that strengthen the idea of one Islamic community and the importance of Islamic solidarity, on both the Chadian and the Saudi sides. For a real understanding of the meaning and the effects of these flows (and hence the connections thus built) in Chadian society, the local context needs to be examined in more detail.

Chad is one of the poorest countries of the world, with a human development index of 0.295, occupying rank 163 of the total of 169 countries (UNDP 2010). The long civil war that affected the country in the 1970s and 1980s coincided with periods of prolonged drought; clan-based management of state resources continues to contribute to political instability today and the ecological fragility of the area remains a constant threat to the existence of many people. Streams of refugees from the Central African Republic and Sudan are adding an additional burden.

The country of old constitutes a transition area between Islamic and Animist and Christian zones, corresponding with the northern and southern parts, respectively. The north has been progressively Islamized over the last seven centuries, partly as a result of the influence of trans-Saharan trade and Arab migration from the east from the fourteenth century onward. European and American Christian missions arrived in Chad from the Central African Republic and Cameroon in the 1930s.

In recent history, religion has been instrumental in struggles for state power by both northern and southern clans. At independence in 1960, a southerner became the country's first president and the majority of the positions in the administration were occupied by southerners who had been educated in French schools that had been boycotted by the majority of Muslims. Another factor contributing to this southern domination was the fact that France had focused in particular on developing the south (*le Tchad utile*). This and the openly negative attitude of the president and his followers led to growing tensions between Muslims and those in power. The northerners seized power in the 1980s after a long period of guerrilla and civil war (during which they did not in fact form a united front). Under Hissene Habre, Chad's first Muslim president, southerners felt that the northerners were progressively occupying their area by installing northern administrators and also with the arrival en masse of Muslim traders and cattle holders fleeing the effects of the drought in the north. Tensions between these Muslim cattle holders and southern Sara-speaking farmers are among the most explosive problems in the south (Arditi 2003) and religion is instrumental in reinforcing "us vs. them" sentiments. The categories are not static however, and successful Muslim traders in southern towns like Sarh are a role model for Sara youngsters and a reason for them to convert. Despite the absence of hard data, most authors would agree that Islam is on the rise in southern Chad (Magnant 1992; Magrin 2001).

In 1990, Habré fell from power and Idriss Deby, one of his former collaborators, took over as president. Although a process of democratization was set into motion, the way the state uses state resources for its own purposes has not fundamentally altered and tensions between the Muslim north and the Christian south continue to exist. Currently, the divisions between the Islamic north and the Christian south are dominant in all thinking on religious, social and political matters, which makes it difficult—and often politically also undesirable—to perceive variation and differentiation within the two camps. The exploitation of the oil fields in the south, which started in 2004, has nourished secessionist sentiments among southerners and the wish to control these sentiments by those in power, and increased surveillance and threats of violence.

Since the 1980s, Chad has (re)intensified its ties with Arab countries economically and religiously/culturally. Chadian imports from the Gulf States have thus increased, while Arab investments in real estate, public works projects, and industries have also grown (Bennafla 2000). In addition to investments, Arab bilateral and multilateral aid has become increasingly important. The Central Market in N'Djamena and the King Faysal University in Chad have, for instance, been financed by Saudi Arabia, with multilateral support mainly being provided through the Islamic Development Bank and the Arab Bank for Economic Development in Africa. Ties with the Arab world have also increased, with growing numbers of Chadians having studied in Sudan, Egypt, or Saudi Arabia, increasing numbers of migrants going to Saudi Arabia but keeping in touch with their families back home and more people from Chad performing the *hadj* (made possible, among other things, by improved means of transport). This intensification of contacts with Arab countries has been accompanied by changing perceptions of Islam and a process of Arabization, which is defined here as an increased cultural orientation toward the Arab world.

The transnational Islamic charities studied in this chapter are part of the increasing connections with the Arab world. They started to arrive in Chad in the mid-1980s when severe drought coincided with the rise to power of Hissene Habre. Internationally, the late 1970s and early 1980s had seen the emergence of the first transnational Islamic NGOs, which was triggered by the war in Afghanistan and made possible financially by the oil boom in the Gulf (Ghandour 2002). These NGOs soon started to work in other parts of Asia and Africa, with Chad being one of the first African countries to receive them. Among the organizations involved was the International Islamic Relief Organisation (IIRO), which was created in 1978 by the Saudi government, and thus can be considered a GONGO[7] rather than a conventional NGO. The arrival of the Sudanese organizations al-Dawa al-Islamiya and International African Relief Agency (IARA) in Chad in the same period can partly be linked to Sudan's foreign policy, which was oriented toward the spread of Arabo-Islamic culture in Africa (Grandin 1993). Al-Dawa al-Islamiya at that time was, however, not only a tool in the hands of Sudan as it was also heavily funded by Colonel Gaddafi and progressively by the Saudis too (Benthall & Bellion-Jourdan 2003:115).

A second wave of international Islamic charities appeared on the scene from the late 1990s onward. In 1993 there was a National Conference in Chad that set in motion a process of (formal) democratization and pacification (Buijtenhuis 1993) and in this context of new beginnings, some well-placed individuals started to lobby in the Gulf States in the hope of attracting Islamic NGOs. They organized meetings and conferences to explain that Chad was a poor country in need of material, social, and cultural support. Saudi Arabia saw an upsurge of new NGOs around this time that were more independent of the Saudi state (but often still administered by members of the Saudi elite) and were seeking an outlet for their funds. They were willing to respond to the Chadian request for aid. Among them were the Al-Makka al-Mukarrama Foundation and the Special Commission of Prince Sultan Ben-Adoul-Aziz for Relief. The Libyan World Islamic Call Association arrived in Chad after the conflict between Chad and Libya over the Aouzou Strip in northern Chad had been settled and Gaddafi's historic visit of reconciliation to N'Djamena in 1998 (cf. Haddad 2000).

There were 11 transnational Islamic NGOs in Chad in 2004, out of a total of approximately 42 transnational NGOs working in Chad at that time. Among them were one Libyan, three Sudanese, one Kuwaiti, and six Saudi organizations. These charities work all over the country, but with an accent on the south, and mainly focus on the building of mosques and wells, education, and supporting orphans. In addition, *da'wa* activities are important, among other things through the distribution of learning materials, the sponsoring of Quranic teachers and preachers who go out into the villages, and the organization of courses in which basic knowledge of Islam is taught.

In theory, the Wahhabist stand of the Arab NGOs opposes the Tidjaniyya form of Islam that is prevalent in Chad. The Superior Islamic Council, which is the highest Muslim authority in the country and is dominated by the Tidjaniyya, has reservations therefore about these new players in the field but it also collaborates with the Islamic NGOs, for instance, in constructing and equipping mosques.[8] The directors of the Islamic NGOs for their part also generally display a collaborative attitude toward the Council. In practice, both the attitudes of the Council and the NGOs are more pragmatic than might be expected if one were to merely look at their ideological stances.[9]

Chadian civil society is not very well-developed (see de Bruijn 2009) with the exception of some NGOs in the Christian south that are working on human rights and agriculture and are mainly supported by Western (Christian) donors. This illustrates how the government tolerates some organization of society in the south but not in the north. This is because the south is seen as a marginal threat to the current government and the clan in power is more afraid of the formation of a strong civil society in the north that could become an important opposition force and therefore blocks it as far as it can. Transnational Islamic NGOs mainly work directly with target groups, such as village communities, sometimes working though local chiefs or local government.

For the ordinary people in the villages and urban neighborhoods, the intervention of Islamic NGOs can offer one of the few opportunities to have a well, a mosque, or medical treatment. The Islamization message may be equally welcome in situations of insecurity. In the south for instance, tensions between mostly Muslim cattle holders and non-Muslim farmers are currently one of the most explosive problems (Arditi 2003). To solve them, the local administration is approached, which, as we have seen, is nowadays usually Muslim. In such a case, it helps if one is also a Muslim and preferential treatment to Muslims is often reported. In this way, Islam enables connections to (local) centers of power. But becoming Muslim is not undertaken for material or political reasons alone. Faith also gives moral support, and having clear rules to follow often adds to a person's feeling of security. A Muslim convert in a southern village, for example, stated that praying five times a day gave him strength and made him feel that he was not alone in the problems he faced. People who had become Muslim in the south also indicated that the ban on alcohol was a good thing in Islam. Whereas before they had spent much of their meager income on alcohol, they now had more money and energy to build something productive. But this does not prevent people from feeling that Islamization leads to cleavages in villages and families, certainly because politics are infiltrating local communities along religious lines. Polarization is being further fueled by the strong competition between the international Christian and Islamic NGOs in southern Chad that are trying to gain converts (Kaag 2007). Transnational Islamic NGOs connect local converts to material means, local centers of power, and transnational webs of morality. At the same time, they also contribute to local processes of polarization between Muslims and non-Muslims and hence to processes of disconnection within local communities.

Transnational Islamic NGOs in Senegal

The transnational Islamic NGOs working in Senegal are largely the same as those operating in Chad, the most important being the Saudi organizations IIRO (International Islamic Relief Organization) and World Association of Muslim Youth (WAMY), the Kuwaiti organization AMA, and, until the recent turmoil and fall of the Libyan government, the Libyan World Islamic Call Society. They find in Senegal a working context that is in many respects very different from that in Chad.

Senegal is also very poor and scores only slightly better than Chad on the Human Development Index. (It is in 17th place on the list of least desirable countries in which to live.) However, the aspect of legal insecurity and state violence that is so obvious in Chad is absent in Senegal. Senegal is a long-standing multiparty democracy; albeit some observers would call it a pseudo-democracy (Diamond 1997). One could indeed question the degree to which parliamentary democracy functions in a country where almost 60% of the adult population is illiterate (CIA 2009) and patron-client relationships form a fundamental part of the makeup of society (Schaffer 1998; Kaag 2001). The fact remains though that changes in government have always been peaceful and conform to the law, freedom of press exists, and there are no major ethnic, religious, or clan-based divisions that serve as polarizing mechanisms, as is the case in Chad.

Ninety percent of the Senegalese population is Muslim. There is a small Christian, mainly Catholic, minority (5%). Most Senegalese Muslims are members of one of the Sufi orders (or brotherhoods) that are prevalent in Senegal. The Tijaniyya is represented in Senegal by the branch of the Sy family in Tivaouane and the Niasse family in Kaolack. The Mouride order is "home-grown" and was founded by Cheikh Ahmadou Bamba at the end of the nineteenth century. The city of Touba is its religious center. While the Tijaniyya are in the majority, the Mouride Sufi order is far more visible and is making a clearer imprint on Senegalese public life than other branches due to the fact that Mourides tend to manifest their religiosity and their allegiance to their marabout (religious leader) more zealously and publicly than others.

There is a small reformist current that mainly evolved from among the student population. From the 1950 onward, young Islamic scholars left Senegal for Morocco, Algeria, Egypt, and Saudi Arabia for theological schooling and after returning to Senegal, some of them established orthodox organizations like the *Union Culturelle Musulmane* (1953), *Al Falah,* and *Jama at Ibadou ar-Rahman* (1978) that oppose the secular character of the state and clientelist relations between the state and the brotherhoods. While many reformist tendencies have their roots at the Cheikh Anta Diop University in Dakar, these organizations became more popular with the urban youth in the 1990s when there was a period of economic crisis and more critical attitudes toward politics developed (Augis 2005). Over the last few decades, more reformist currents within the Sufi orders have also gained influence, as is exemplified by organizations such as the Hizbut Tarqiyya in the Mouride order that takes a critical stance against the all-powerful positions of the marabouts and refutes hereditary claims to maraboutic power and the Moustarchardine movement within the Sy branch of the Tidjane brotherhood (Gueye & Seck 2011).

The different religious groups have always lived in harmony.[10] This is to a certain extent related to another important characteristic of the Senegalese context which is the long-standing symbiotic relationship between the state and the Sufi orders. Dating from colonial times, this has continued since independence. The state elite needs the political support of the brotherhoods to stay in power, and gives the brotherhoods economic and political advantages in return. For a long time this has helped to create stability in the country. The worldly order and the religious sphere remained distinct domains but mutually supported each other to the benefit of both political and religious elites. The last few years, however, have seen a growing interpenetration of the religious and political domain by religious actors actively engaging in politics, and politicians openly using religious registers (Gueye & Seck 2011). The president who came to office in 2000 is a clear example of the latter development. Unlike the former presidents Senghor (who was a Catholic) and Diouf (who is a Muslim but never declared this publicly and came across as more of a technocratic ruler), Abdoulaye Wade, who reigned until 2012, clearly portrayed himself as a Mouride *taalibe* (pupil, follower) and used this quality politically. Religious leaders have, in their turn, started political parties and invested in existing parties (Gervasoni & Gueye 2005; Gueye & Seck 2011). This has contributed to Islam becoming more central in public debates and has led to lively discussions in the public sphere about "proper" Islam, the role of Islam in society, and the secular character of the Senegalese state (Kaag 2011a).

In contrast to Chad, Senegal has a strong and diverse civil society. In 2008, there were 469 registered NGOs, of which 295 were national and 174 were international organizations (Direction du Développement Communautaire 2008). In addition, there are many religious associations, rotating credit and savings groups, neighborhood self-help groups, professional associations, and development and environmental associations.

How do transnational Islamic relief organizations fit into this context? First of all, it is clear that they are less visible in Dakar than in N'Djamena, as they blend in with the multitude of signs indicating the headquarters of international, national, and local NGOs and associations.

While it would be logical that transnational Islamic NGOs in Senegal worked with local NGOs and associations as there are so many possible partners, in reality most organizations claim to work mainly with individuals who come with requests for assistance, which is their normal approach. In view of the level and tradition of organization in Senegal, a lack of fit—that is, of connectivity—can be discerned here. When asked why local associations and NGOs do not come with requests, WAMY's director replied that it was probably because these organizations have another signature and their own ways of securing funds. This is, of course, a logical argument in a Sufi-dominated Islamic landscape. It is remarkable, however, that this argument holding that differences between the Sufi and the more Salafi form of Islam propagated by most Arab NGOs blocks collaboration between the two parties does not hold true for the Sufi leaders and the Arab NGOs. WAMY, for instance, organized a medical caravan in Touba at the request of the current general Caliph of the Mouride order. The director of WAMY proudly declared that he was also in touch with the Caliph of the Tidjaniyya in Tivaouane but that this has not yet yielded any concrete projects.

The World Islamic Call Society (WICS) also organized a medical caravan in Touba at the request of the Caliph of the Mourides. Interestingly, the Chief of Cabinet of WICS's director is a member of the Touba maraboutic family. The Libyan attitude toward Sufism has been more positive than that associated with the Arab peninsula, as Gaddafi always stressed that he was in favor of Sufism, considering it a form of Islam specially adapted to Africa and its culture. This is also stressed by WICS in Senegal. It appears, however, that this no longer offers the Libyan organization a more favorable position, as the Arab NGOs also appear open to collaborating with Sufi actors and vice versa.

There is also collaboration with the Senegalese state, for example in the field of health care and education. President Wade has actively pursued a strategy to strengthen links with the Arab world and managed to host the 11th Organization of Islamic Conference s Summit in Dakar in 2008. The director of one of the Arab NGOs, who had previously worked in Mali and Niger, welcomed the active attitude and openness of the Senegalese state to collaboration but complained that "the State is too secular," which in his view causes difficulties in obtaining land for projects and the like.

Apart from the state and the Sufi leaders, WAMY and AMAI also collaborate with the university and more reformist (student) organizations such as the *Union de la Jeunesse Musulmane du Sénégal*. WAMY, for instance, donated computers to the Arab Department of Cheikh Anta Diop University in Dakar and a large number of books to help increase students' knowledge of Arab culture. Both WAMY and AMAI also organize seminars for students and imams. The Saudi organization International Islamic Relief Agency works less in the university environment and is more engaged in activities like building mosques and wells, and providing care for orphans.

At the village level, connections between local communities and transnational Islamic NGOs are being facilitated by religious leaders too. The Imam of the Grand Mosque in Kaolack, for instance, encouraged local religious leaders in the region to put in requests for mosques to a transnational Islamic NGO with its headquarters in Dakar. This call was taken up in the village of Kaymor by a local *arabisant*,[11] and he proudly showed me the new mosque when I visited the village in 2010. He claimed that the fact that he was an *arabisant* and knew former fellow students who worked at the NGO's Senegalese head office had facilitated contacts with the organization. The village had always been known as a pagan village as it had been islamized quite late, and as an commercial center, it was felt that moral values were taken less seriously here. With the arrival of the new mosque, Kaymor could finally boast its religious importance and demonstrate its commercial and administrational clout. But whereas the building of a mosque in southern Chad may be seen as a political act of territorial occupation in a religiously polarized landscape, a mosque in a Senegalese village such as Kaymor is primarily welcomed as an infrastructural asset and has no political connotation beyond a possible local political one, as, for instance, in the case of a local leader wanting to increase his influence by donating or facilitating its building.

AMA is the only transnational Islamic relief organization with its national headquarters in the south of the country, in Ziguinchor, the capital of the Casamance. This is a region that is very different from the rest of Senegal: it has been engaged in a long (guerrilla) war of secession characterized by a great deal of insecurity for the local population. The additional fact that there is no Muslim majority but cohabiting Muslim and Christian/Animist groups makes the situation comparable to southern Chad, although the clear link between Islam and power is lacking in the Casamance. AMA started working in the region in 1991, a year after war broke out (Marut 2002). I have not yet been able to extend fieldwork to the Casamance but it will be interesting to see what kind of dynamics are being triggered by AMA working in this area.

Connectivities Compared

The interventions of transnational Islamic NGOs seem to be of more importance socially in Chad than in Senegal, as poverty and suppression (social violence and structural violence from the state) are harsher there while, in general, aid interventions are fewer. This means that (promises of) interventions are seen as important by the population as they are one of the few opportunities to receive material and/or moral support. In addition, transnational Islamic NGOs seem to have more political clout in Chad. They are intervening in a political context where Arab identity and Islam are decisive for being in power or being marginal,[12] and these organizations are thus associated with access to power and with the reigning elite. Whether they want it or not, they are also becoming part of the polarization, violence, and political dynamics associated with the religious divide, which has repercussions for the social roles they are able to play.

Transnational Islamic NGOs in Senegal are few in number. Islam per se is less an issue there but the kind of Islam, however, is an issue and in this dynamic, reformist strands that are close to the reformist agenda of the transnational Islamic NGOs are a minority but are increasingly visible in the public sphere. They may use (moral, financial, or political) support by these organizations but do not need to, as they also have a strong local stance and there are other local organizations with which one can associate when the need is felt or when this seems favorable. There is, however, also more rivalry between groups, in which help from transnational organizations may be useful. People (target groups) in Senegal seem to have more options than in Chad and are less dependent on outside support from transnational Islamic charities. What is interesting is that both state actors and religious leaders from the Islamic Sufi orders are increasingly interested in collaboration. The resources brought in by these charities are used to win, co-opt, and/or satisfy their own clientele. It appears that the Arab charities are less able to put their stamp on their work or to generate publicity for its own sake and on their own account.[13] This all means that transnational Islamic NGOs in Senegal are much more marginal from a political and social perspective than in Chad. Whereas in Chad the environment seems to facilitate transnational Islamic NGOs in their work both socially and politically, the environment in Senegal is such that they have to content themselves with a more dependent and marginal position.

In terms of connections, it can be concluded that the Chadian case shows a strong level of connectivity in that the characteristics of the Chadian context facilitate connections between Chadian target groups and transnational Islamic NGOs, and the building of a transnational moral network such as that aimed for by the latter. The Senegalese case shows connectivity at certain points, in particular where it concerns the state and religious elites who want to collaborate with these organizations. A further comparison draws attention to the content of the connections established and the center of gravity of the networks built. Whereas the moral network in Chad appears to be transnational with reference to the Arab world and the wider *umma* is seen as important, transnational connections in Senegal seem to feed into networks with a more localized center of gravity. These factors have repercussions for processes of local disconnect. There are clear signs of these processes in Chad where people are becoming involved in the activities of transnational Islamic NGOs, in particular when answering their *da'wa* in southern Chad, and this is creating oppositions and tensions in families and local communities. In Senegal, this seems to be less the case.

Concluding Remarks

This contribution has aimed to show how transnational Islamic NGOs from the Gulf region are offering not only relief but also new ways of connecting to the *Umma*. They are inspired by specific understandings of Islamic solidarity and are weaving transnational moral networks by linking Arab donors to African target groups with messages focusing on the propagation of what they consider the right Islamic practices. Local contexts and their connectivity play an important role in determining the character and importance of the connection, as was shown by comparing the cases of Chad and Senegal. These examples demonstrate that transnational Islamic NGOs are not the mighty intruders forcing defenseless Africans into a radical form of Islam but ordinary organizations dealing with the local contexts in which they are working, circumstances that, for a large part, determine their work modalities, the responses they receive, and from whom. This illustrates how transnational Islam is being coshaped by African agency that is influencing new global connections as much as outsiders' agency does. Using connections as an analytical tool has allowed an accent on flow and transmission between two points of connectivity, and has enabled the presence of transnational Islamic NGOs from the Arab world in Africa to be viewed as a two-way process in which the two connecting points both have an input and are coshaping the (provisional) outcome. Moral networks are being weaved, in which material aid and moral messages have been the input of the Islamic NGOs, while the local side provides blessing in return. At a more practical level, the fit between what the NGO has to offer and the local context largely determines the outcome, that is, the content, the meaning, and the connections established. In the two cases described, the connection is having a redistributive effect in the sense that material means are being transferred from rich Muslims in the Arab world to poor Muslims in Africa. The difference would seem to lie in the fact that the connection in Chad has a clear disciplining effect, in the sense that the moral message has more impact socially and politically than in Senegal. Further research at the grassroots level in the local communities in which these organizations are working will yield further in-depth knowledge of the ways the factors identified in the political context, the characteristics of poverty, and the ways in which "development" and "civil society" are organized influence interaction between these organizations and their African target groups and will flesh out the consequences for local and transnational (dis)connections.

Notes

1. *Zakat* is fixed at one-fortieth of one's annual assets. The assets that have to be counted are determined by Islamic jurisprudence. Muslims with an income below a certain threshold (*nisab*) do not have to contribute at all. It should be noted that after 9/11 when international money transfers to Islamic organizations came under greater scrutiny, transnational Islamic NGOs increasingly used *waqf* (a sort of Islamic endowment, the proceeds of which are used for religious or charitable purposes) as a source of funding (Petersen 2011).

2. As with many issues in Islam, *zakat* is subject to multiple interpretations. It is, for instance, a contentious issue whether *zakat* can be given to one's own family. In this chapter, I am not so interested in theoretical and theological debates but in practices of groups and people. Religious specialists would perhaps say that the money that Murid migrants transfer to their families cannot be *zakat* but if the migrants themselves call it *zakat*, I consider that it is.

3. Other Islamic NGOs, such as the UK-based Islamic Relief, take a more neutral stance and provide aid for non-Muslims and Muslims alike, without having *da'wa* as a strategy. See also Petersen (2011) who makes a distinction between transnational Islamic NGOs with a sacralized aid ideology and those with a secularized aid ideology.

4. As was strongly felt, for example, by my Chadian research assistant who considered the connotation to be that Arabs are naturally better than Africans and that Africans would evidently want to be like them.

5. More recently, a more positive perspective on Islamic NGOs has been emerging in the framework of renewed interest in faith-based organizations (FBOs). Interestingly, these two opposite perspectives have their roots in the same recognition that culture matters in today's global era. Interpreted negatively, this could lead to the kind of "clash of civilization" thinking that is so apparent in the War on Terror discourse. Interpreted positively, this leads to the idea that FBOs, including Islamic NGOs, are often better equipped to offer relief than others because of a supposed cultural proximity to their target groups (Kaag & Saint-Lary 2011).

6. Analysis is based on a literature review, fieldwork in Chad in 2004 (N'Djamena and the Moundou Region in southern Chad), and interviews held in Senegal between 2009 and 2011 (Dakar and Kaymor in central Senegal).
7. Government-Operated Non-Governmental Organization, see Naim (2007).
8. However, the Council is not pleased that the activities of these NGOs fall under the supervision of the Ministry of Planning. As such, it has no control over them. They see this as unjustified because the Council's task is to supervise all Islamic activities in the country (see al-Mouna 2004).
9. Contributing to this quite open attitude is the fact that the directors of these organizations are mostly expats with extensive international experience. The Director of al-Makka al-Mukarrama, for example, is a Saudi who served in Bosnia before moving to Chad, and the Director of AMA is a Moroccan who used to work in Burkina Faso.
10. The Senegalese are proud of the fact that Christians and Muslims have always cohabited peacefully, and that their first president, Senghor, was a Christian but led an Islamic country without any religious tensions or problems with the Islamic leaders.
11. A specialist in Arab language and culture. Many *arabisants* in Senegal have studied in Arab countries.
12. Even though, ultimately, ethnic adherence is also important.
13. As already mentioned, the situation in the Casamance is different from the rest of Senegal, and may be more comparable to Chad. There is no Muslim majority but rebellion, war and violence, and strong competition between Islamic and Christian NGOS. What is different in comparison with Chad is that the strong association of Islam with power and of not being a Muslim with powerlessness is lacking.

References

Arditi, C. (2003) "Les Violences Ordinaires ont une Histoire: Le Cas du Tchad," *Politique Africaine* 91: 51–67.

Augis, E. (2005) "Dakar's Sunnite Women: The Politics of Person," in: M. Gomez-Perez (ed.), *L'Islam politique au sud du Sahara. Identities, discours et enjeux*. Paris: Karthalia, pp. 309–326.

Bennafla, K. (2000) "Tchad: L'Appel des Sirènes Arabo-islamiques," *Autrepart* 16: 67–86.

Benthall, J. & J. Bellion-Jourdan (2003) *The Charitable Crescent. Politics of Aid in the Muslim World*. London & New York: I.B. Tauris.

Bornstein, E. (2002) "Developing Faith: Theologies of Economic Development in Zimbabwe," *Journal of Religion in Africa* 32(1): 4–31.

Bornstein, E. (2003) *The Spirit of Development. Protestant NGOs, Morality, and Economics in Zimbabwe*. New York & London: Routledge.

Bruijn, M. de (2009) "The Impossibility of Civil Organizations in Post War Chad," in: A. Bellagamba & G. Klute (eds.), *Beside the State. Emergent Powers in Contemporary Africa*. Köln: Rüdiger Koppe Verlag, pp. 89–105.

Buijtenhuis, R. (1993) *La conférence nationale souveraine du Tchad: Un essai d'histoire immédiate.* Paris: Karthala.

CIA (2009) *World Factbook: Senegal.* Washington, DC: Central Intelligence Agency.

Diamond, L. J. (1997) *Prospects for Democratic Development in Africa. Essays in Public Policy No 74.* Stanford, CA: Hoover Institution, Stanford University.

Direction du Développement Communautaire (2008) "Contribution de la Direction du Développement Communautaire à l'atelier du CENTIF sur le financement du terrorisme du 18 au 20 décembre 2008." Republique du Sénégal, Ministère de la Famille, de la Solidarité Nationale de l'Entrepreunariat Féminin et de la Micro Finance.

Gervasoni, O. & C. Gueye (2005) "La confrérie mouride au centre de la vie politique sénégalaise. Le «sopi» inaugure-t-il un nouveau paradigme?," in: M. Gomez-Perez (ed.), *L'Islam politique au Sud du Sahara.* Paris: Karthala, pp. 621–639.

Ghandour, A-R. (2002) *Jihad Humanitare. Enquête sur les ONG Islamiques.* Paris: Flammarion.

Grandin, N. (1993) "Al Merkaz al-islami al-afriqi bi'l-Khartoum. La République du Soudan et la Propagation de l'Islam en Afrique Noire (1977–1991)," in: R. Otayek (ed.), *Le Radicalisme Islamique au Sud du Sahara.* Paris: Karthala, pp. 75–95.

Gueye, C. & A. Seck (2011) "Islam et politique au Sénégal: Logique d'articulation et de co-production," in: M. Kaag (ed.), *Islam et engagements au Sénégal.* Leiden: African Studies Centre.

Haddad, S. (2000) "La Politique Africaine de la Libye: de la Tentation Impériale à la Stratégie Unitaire," *Monde Arabe. Maghreb-Machrek* 170: 29–38.

Hunwick, J. (1997) "Sub-Saharan Africa and the Wider World of Islam: Historical and Contemporary Perspectives," in: D. Westerlund & E. E. Rosander (eds), *African Islam and Islam in Africa. Encounters between Sufis and Islamists.* London: Hurst & Co., pp. 28–54.

Kaag, M. (2001) *Usage foncier et dynamique sociale au Sénégal rural.* Amsterdam: Rozenberg Publishers.

Kaag, M. (2007) "Aid, *Umma* and Politics: Transnational Islamic NGOs in Chad," in: R. Otayek & B. Soares (eds.), *Muslim Politics in Africa.* New York: Palgrave Macmillan, pp. 85–102.

Kaag, M. (2008) "Transnational Islamic NGOs in Chad: Islamic Solidarity in the Age of Neoliberalism." *Africa Today* 54(3): 3–18.

Kaag, M. (2010) "L'opera delle ONG islamiche transnationalli: Ciad e Senega a confronto (The Work of Transnational Islamic NGOs in Africa: Chad and Senegal Compared)," *Afriche e Orienti* (2): 102–115.

Kaag, M. (ed.) (2011a) *Islam et engagements au Sénégal.* Leiden: African Studies Centre.

Kaag, M. (2011b) "Connecting to the *Umma* through Islamic Relief: Transnational Islamic NGOs in Chad," Special Issue, Translocal Development, Development Corridors and Development Chains, *International Development Planning Review* 33(4): 463–474.

Kaag, M. (2011c) "Zakat building transnational connections with and within Africa," Reflections on the basis of research on transnational Islamic NGOs and the Senegalese diaspora", Paper presented at the International Workshop "Zakat in a Comparative Light". Geneva, January 17–18, 2011.

Kaag, M. & M. Saint-Lary (2011) "The New Visibility of Religion in the Development Arena. Christian and Muslim Elites' Engagement with Public Policies in Africa," in: M. Kaag & M. Saint-Lary (eds.), Special Issue Religious Elites in the Development Arena, *APAD Bulletin* (33): 23–37.

Magnant, J-P. (ed.) (1992) *L'Islam au Tchad*. Bordeaux: CEAN.

Magrin, G. (2001) *Le sud du Tchad en mutation. Des champs de coton aux sirènes de l'or noir*. Saint-Maur-des-Fossés: Sépia.

Marut, J-C. (2002) "Les particularismes au risque de l'Islam dans le conflit casamancais," in: CEAN, *Islams d'Afrique, entre le local et le global*. Paris: Karthala.

Al-Mouna (2004) "Dossier: les religions telles qu'elles sont vécues par els Tchadiens," *Carrefour*, no. 28 (July-August).

Naim, M. (2007) "Democracy's Dangerous Impostors," *The Washington Post*, April 20.

Petersen, M. J. (2011) "For Humanity or for the *Umma*? Ideologies of Aid in Four Transnational Muslim NGOs." Ph.D. Thesis, University of Copenhagen.

Rosander, E. E. (1997) "Introduction: The Islamization of 'Tradition' and 'Modernity'," in: D. Westerlund & E. E. Rosander (eds.), *African Islam and Islam in Africa. Encounters between Sufis and Islamists*. London: Hurst & Co, pp. 2–27.

Schaffer, F. C. (1998) *Democracy in Translation: Understanding Politics in an Unfamiliar Culture*. Ithaca, NY: Cornell University Press.

Soares, B. F. (2005) *Islam and the Prayer Economy. History and Authority in a Malian Town*. Ann Arbor, MI: University of Michigan Press.

UNDP (2010), *Human Development Report 2010*. New York: United Nations Development Programme.

Weiss, H. (2002) *Social Welfare in Muslim Societies in Africa*. Uppsala: Nordiska Afrikainstitutet.

Westerlund, D. & E. Evers Rosander (eds.) (1997) *African Islam and Islam in Africa. Encounters between Sufis and Islamists*. London: Hurst & Co.

CHAPTER 10

Love Therapy: A Brazilian Pentecostal (Dis)connection in Maputo

Linda van de Kamp

Introduction

The growth of Pentecostal churches in Africa[1] has coincided with cultural, social, and economic changes that have altered the meaning of sexuality and love amongst younger generations (Geissler & Prince 2007a, 2007b; Soothill 2007; Bochow 2008, 2010; Cole 2010; Frahm-Arp 2010; van de Kamp 2011a). With an absence of role models amid changing interpretations of love and relating, the teachings of Pentecostal pastors on these issues have been taken as examples. Pentecostals are known for their open therapeutic discourse on sexuality and relationships, and taboos concerning sexuality, love, marital relations, infidelity, and domestic violence no longer exist. According to Pentecostals, people need to discuss these subjects as a prerequisite to resolving their problems (cf. Spronk 2006: 227–229). Premarriage counseling sessions, workshops, and group meetings for married couples are thus being organized to consider love, sexuality, and marriage. The role of religion in education about family, marriage, and sexuality is not new and there is a large body of literature on colonial Christian missions and the shaping of new identities in relation to work, health, the family, sexuality, and marriage (e.g., Comaroff & Comaroff 1991, 1997; Ranger 1992).[2] Nevertheless, the era of AIDS prevention and treatment programs has renewed the Christian public discourse on sexuality and marriage due to the increasing influence of faith-based organizations (FBOs) in AIDS programs (Becker & Geissler 2007, 2009; Burchardt 2009, 2011; Prince *et al.* 2009; van Dijk 2010).

The Christianization of public discourse on sexuality and relationships is increasingly being shaped by new transnational Christian connections in Africa. Brazilian Pentecostals are establishing South-South connections in Africa, such as in Maputo, the capital of Mozambique. Every Saturday evening, the Brazilian Universal Church of the Kingdom of God (Universal Church)[3] organizes *terapia do amor* (Therapy of Love) in several of its branches there.[4] The therapy[5] is a public form of counseling that resembles a church service but is directed at producing new discourses and realities of love, sexuality, and marriage. Thousands of people, mostly young women aged 15–35, participate every week. The heart of the therapy lies in learning how to express and create love publicly by connecting participants to a Brazilian transnational discourse on "romantic love" that aims at transcribing a new morality of love, sexuality, and marriage into the lives of Maputo's citizens. An illustration of this is when the pastor asks couples to come forward and stand in front of the podium. He then says: "Now apologize to each other for the mistakes you have committed towards each other." Romantic music plays and the pastor and his wife stand on the stage as he continues: "Embrace each other, kiss your partner and say: 'I love you.'" The pastor and his wife do the same and start praying for the couples in front of them.[6] In the Mozambican setting, as in other African societies (cf. Spronk 2002 for Kenya), it is unusual and often considered shocking to openly express affection for your partner in the presence of others and to talk about sexuality in a public forum.

This chapter examines this public training of the body in ways of love, such as embracing and kissing, in relation to the changing practices of love and new gender roles in Maputo.[7] I examine Brazilian Pentecostal counseling sessions on love and sexuality as a set of "connecting techniques" where the *terapia do amor* serves as a key example. In line with the arguments running through this current volume, connecting techniques are specific forms of linking that people consider to open up new life options, such as the invention of new bodily modes and relationships. In the case of the therapy, the connecting techniques have two important meanings. First, the value of the Brazilian Pentecostal connections for Mozambican urban women is intrinsically related to its transnational aspects. The transnational Pentecostal bridge allows for disconnecting from existing forms of relating and learning about alternative ways of being and relating. Second, it appears that the embodiment of specific constructs, tools, or techniques (Foucault 1988) produces love and successful relations. To connect to alternative forms of love and marriage and to disconnect from older ones, the body plays a central role in realizing connections and effectuating sociocultural change. The last part of this chapter describes how the new modes of bridging and bonding through the embodiment of Brazilian Pentecostal techniques are also leading to insecure feelings and relationships.

Love, Sexuality, and Marriage in Maputo

The growing literature on the anthropology of romantic love (Spronk 2002, 2006; Etnofoor 2006; Hirsch & Wardlow 2006; Bochow 2008, 2010) illustrates how young people around the world are talking about the importance of emotional intimacy in the creation of affective bonds and marital ties. Emotional experiences, such as passion and pleasure, have become fundamental to modern love. What is striking in this respect is that courtship and conjugal relationships are demonstrations of modern individuality rather than relationships that create obligations between kin groups as well as between individuals (Giddens 1992; Illouz 1997). A shift has occurred from "arranged marriages" to "marriages of love." Love has come to serve as the ultimate expression of a progressive relationship. The apparent "global ideological shift in marital ideas" (Hirsch & Wardlow 2006: 2) is related to transformations of socioeconomic structures as well as to cultural globalization (Lindholm 2006). Marriage alliances in many societies, including Mozambique (Feliciano 1998: 249–267), used to be essential to economic and political survival and marital relations were negotiated by powerful elders who secured the advance of their family and community. In these circumstances, marriage was a necessity and a duty, and romantic attraction was a potential threat to the stability of the extended family (Lindholm 2006: 11). However, the changing organization of production and consumption has led to more powerful positions for young people and the formation of nuclear families, a trend that is influencing marital relations and ideas (Collier 1997; Illouz 1997).

Most of the available literature on sexuality and marriage in Mozambique is recent and dates from after the time when the AIDS epidemic started to be felt in the 1990s. As more women are infected with HIV than men—about 60% of all HIV-positive adults are female, particularly in the 15–24 age group (UNAIDS/WHO 2008: 4–5)—most of the studies are on women, gender, and youth. Feminist-oriented works predominantly demonstrate how, due to patriarchal structures, men control women's sexuality (Cruz e Silva *et al.* 2007; Osório & Cruz e Silva 2008; Osório 1997; Santos & Arthur 1994; cf. Loforte 2003). As men engage in multiple relations and polygamous marriages, women run a higher risk of becoming HIV-positive because using a condom in marriage or in a steady relationship is considered problematic: the condom signals distrust. Women's socialization into sexuality, both at home and at school, occurs with reference to marriage and reproduction. Girls are warned against getting pregnant outside of marriage and within marriage sexuality is about becoming pregnant and having children. Boys and men have more freedom to experiment with sexuality and have the right to demand sex from women and experience its pleasures. Studies have concluded that sexuality is marked by inequality between women and men, with women having little say on sexual issues and exercising limited power over their bodies and sexual desires.

A younger generation of scholars that examined narratives, practices, and views on sex amongst the youth in Maputo (Hawkins *et al.* 2005; Karlyn 2005; Gune 2008; Manuel 2008, 2011) has demonstrated that while there are relations of power between men and women, youth are actively involved in developing new sexual practices and relations. These scholars describe how young women are challenging patriarchal rules and redefining their roles in relation to men (cf. Loforte 2003: 201–223). For example, women are developing new social and sexual identities by engaging in multiple occasional sexual partnerships and using them to improve their socioeconomic position (cf. Thornton 2009). And a growing group of upwardly mobile women is no longer dependent on marriage for economic survival (cf. Penvenne 1997) and their role in biological reproduction is not decisive in establishing their position and identity either. They are exploring new sociocultural domains and lifestyles, including new forms of love and relating, that are part of the "global ideological shift in marital ideas" (Hirsch & Wardlow 2006: 26).

Maputo youth generally distinguish three types of sexual relationships: the steady relationship (*namoro*/courtship) in which sexuality involves love; the occasional one (*saca-cena* or a one-night stand and *pito/pita* or having a regular male/female sex partner) where love is not an issue; and the *ficar* (to stay) relationship that is somewhere between the other two. The meaning of *ficar* in the context of an amorous relationship has been adopted from Brazilian *telenovelas* (soaps) that are broadcast on Mozambican television.[8] In occasional relationships, which may last years or just one night and vary from an exchange of kisses to having sex, there is no bond between the partners. In the words of one of Manuel's (2008: 42) young male research participants: "We do not take her home," "There is no intimacy," "I do not value her like I value a *namorada* [girlfriend]." One young woman described a *pito* as "the one with who you cheat on your *namorado* [boyfriend]."

Namoro on the other hand is about respect, affection, and a bond. This is a serious relationship. In the case of *namoro*, parents and a wider circle of kin are involved to a greater or lesser degree[9] and it is common to hear talk about love (*amor*). For the young upwardly mobile women I met, both Pentecostal and non-Pentecostal, love meant that they and their boyfriends were often together, shared future dreams, and supported each other. Love was similar to trust and no secrets should exist (cf. Manuel 2008). There are several rules attached to a *namoro*. For example, *namorados* are expected to go out together, though this applies more to the woman than the man. Often, the *namoro* relationship is consolidated with "real sex": sex demonstrates love and trust. By having real sex there is intimate contact with the partner because "blood" is exchanged. Here, blood is a symbol for what partners give each other in the exchange of bodily fluids, which proves love.

The upwardly mobile people, mostly women, I met who frequented the *terapia do amor* were all concerned with *namoro* in various ways. They were often not able to establish a *namoro* but could be in a *ficar* relationship. This relationship is more than occasional but is not steady. It may potentially develop into a serious relationship but women felt that it was difficult to change a *ficar* situation into a serious liaison. If they showed their wish for a steadier relationship, they risked their boyfriend leaving or disappearing suddenly. In the case of an evolving *namoro*, the man might disappear as soon as the girlfriend's family became involved. Women also complained that they felt obliged to have sex with their boyfriends: they had to show that they were serious about the relationship but they were not sure whether their boyfriend was as serious as they were. The men said they wanted to *namorar* but in fact wanted to have sex, with the social discourse about occasional sex being mostly negative. Some women were afraid that their *namorado* had a *pita* or *saca-cena*. In addition, they visited the counseling sessions given by Pentecostal pastors because they wanted a different kind of *namoro*. Sometimes they and their boyfriends differed in their opinions about the meaning of love. To some men, a *namoro* meant that they should control their girlfriend as that was proof of their love. But the girl would experience the control as a form of her partner's distrust of her. Men spoke about similar experiences and referred to *estar engarrafado* (being bottled). This is a male expression suggesting that their girlfriends are controlling them and they do everything their girlfriend wants.[10]

Another frequent problem is the role of kin in relationships. Traditionally, sexuality is regarded as part of the transition from childhood to adulthood, in which kin play a central role (Loforte 2003: 201–223). The transition was always carefully guarded by counseling sexuality through initiation rites organized by kin and/or respected elders. However, these rites are disappearing, particularly in southern Mozambique where the influence of colonialism, missionaries, and socialism was strongest.[11] Moreover, several people, including *curandeiros* (local healers), said that destroyed kinship structures, often as a result of the civil war (1976–1992), had impeded the transfer of knowledge from one generation to the next and that young upwardly mobile urban couples were more independent from kin due to their education and jobs.

As part of this process, young people, including young converts, want to control their sexuality individually and without interference from kin. Young people define adulthood as independence from parents, which they express through their control of their own sexual encounters, and thus of their bodies (cf. Manuel 2008: 38). Yet while *namorados* find that they are adults and responsible for their own deeds, this is not always an opinion shared by their kin who may intervene in the relationship and try to influence the *namoro* and any marriage arrangements. This results in conflicts because interventions by parents or kin come too late or make no sense to the *namorados* (Karlyn 2004; Paulo 2005). Part of the problem is related to the fact that, according to local customs, parents do not talk about sexuality and marriage with their children (Loforte 2003: 211; Paulo 2005; Cruz e Silva *et al.* 2007: 93, 120–122). For example, Elena (a university student)[12] told me that topics related to courtship were taboo and not a subject of conversations at home. She had never talked to her mother about it, "and now I am 26!."[13] Today, youth are finding out about sexuality from friends, at school and from television, while custom prescribes that kin are in control of sexual education. Elena said: "We learn with friends, we learn about it outside."

All the particular dynamics at the levels of interaction in the field of sexuality and relating were important reasons for frequenting Pentecostal counseling sessions. Women who participated in these meetings were uncertain if they wanted to get married because of all the risks involved, such as domestic violence, infidelity, and HIV/AIDS. Staying single and/or engaging in occasional relationships gave them more freedom and power but if they were to be respected, they were expected to marry and become mothers. Their engagement with Brazilian counseling practices were thus shaped by a mix of ambiguous feelings and experiences as well as by the chances to shape new reproductive lifestyles.

Brazilian Pentecostal Counseling: "Terapia do Amor"

Mozambican churches do not in general openly organize special teachings on issues of sexuality, love, courtship, and marriage, particularly not for young, unmarried people. Most churches, such as African Independent Churches, the Catholic Church, and older Protestant churches, have women's groups where *coisas do lar* (issues related to the home and marriage) are discussed (cf. Igreja & Lambranca 2009: 278–279 for central Mozambique; cf. Soothill 2007: 104–108 for Ghana), but only married women are allowed to participate. Love and sexuality are not topics to be discussed with unmarried persons. Married women first asked me whether I was married (and whether I thus knew about sexuality and marriage) before they would talk about them with me. On the contrary, most of the Pentecostals pastors are open about these topics toward youth in the Brazilian churches or churches with Brazilian links (e.g., the Assemblies of God), which was an important reason for women to attend these churches. For example, Ana (a 21-year-old high-school student) said: "I went to the Universal Church because there they talk about our problems [such as unfaithful men]. There they speak about courtship."

While pastors may preach on love, sexuality, and marriage during services, this normally happens at meetings in youth groups and/or special workshops. These meetings often take the form of a discussion group with the aim of counseling Mozambicans about love. Counseling is a form of behavioral change and offers practical advice. One receives advice and guidance from a knowledgeable person on, in this case, how to love (Burchardt 2009; Nguyen 2009; van Dijk 2010). In Pentecostal booklets and the *Folha Universal*, articles and columns with advice are published on how to change partners and stay happily married.[14] They all involve a particular therapeutic discourse on affective relationships (Illouz 1997: 198).

Counseling practices are generally performed publicly and only the premarriage part takes place in private. It consists of one or two meetings between the couple and the pastor and his wife who teach about marriage. Counseling practices generally are very open because sexuality has always been related to private domains and, according to Pentecostals, has been silenced. The focus is now on more openness. Taboos relating to love, sexuality, and marriage need to be addressed and Mozambicans have to learn to talk about these subjects as a prerequisite to resolving their problems. Workshops, discussions, and group meetings are thus being organized. In the Brazilian Renewed Baptist Church, debating sessions for youth were set up on *namorar* (courting). Youth from the Maná Pentecostal Church met church leaders every Saturday to discuss a subject such as local forms of marriage (van de Kamp & van Dijk 2010: 130–133). At one of the churches of the Assemblies of God, a group for married women aged between 18 and 34 was set up in 2001 with the assistance of Brazilian missionaries.

These counseling activities were characterized by a conversational technique. Though the leaders generally made it clear what kind of sexual and marital lives converts should live—abstaining from sex before marriage, fidelity, caring for one's partner, enjoying sexuality within marriage, and keeping a critical distance from "tradition"—there was room for debate, doubts, questions, and opposing views. In contrast, the *terapia do amor* at the Universal Church is a much more fixed and controlled "treatment" and route to success. Romantic love is created through prescribed formats and techniques and there is no place for doubt or uncertainty. Moreover, the relationship of confidence that usually exists between counselors and clients, and which could be seen in the discussion groups, was impossible to recreate because hundreds of persons gathered for the *terapia do amor* sessions.

The Universal Church organizes its *terapia do amor* in the *Templo da Fé* (Temple of Faith) and in a few other buildings in Maputo on Saturday evenings and between 1,000 and 2,000 people from all over the city, most of them youngsters, attend every week. Usually the therapy starts with the pastor telling people to put their hands on their hearts, something that happens in almost every church service as well. Gospel music then plays to lift people up and invoke the Holy Spirit. This is followed by prayers when people lift up their hands to God in surrender and talk to Him. They cry or shout their intimate problems (*problemas sentimentais*). There is always a sermon on a topic such as how to find a good partner, how to trust each other, whether anal sex is allowed, what love is, or what marriage should feel like. Participation is important and the pastors ensure the audience is active and interacts with them. Everyone has to look up Bible passages, repeat words or complete phrases, give the pastor answers, look him in the eye, and talk to their neighbors. The pastors are lively, enact sketches and call assistants forward to play roles, using variations in their voices, making jokes and creating fear.

Special performances are also held. On one occasion,[15] a huge heart stood at the front of the church and everyone had to walk under it while the pastor and his assistants prayed that God would provide them with luck in their love lives. They had to form a row of women and a row of men and were paired up to walk through the heart together. Then all the married couples had to come forward, embrace each other and apologize if they had quarreled or if there had been a lack of understanding between them. The pastor too embraced his wife and both prayed for the couples. The whole performance was accompanied by music and sound effects and it was clear that one should engage fully in the therapy in order to achieve results.

Denouncing destructive powers is an important element in the therapy. For example, at the start of one session, the pastor explained why jealousy and infidelity characterize so many relationships. Sometimes family members engage in witchcraft because they dislike their daughter-in-law, the pastor explained.[16] Or a woman consults a *curandeiro* to receive herbs that will make a man who already belongs to another woman change his mind and be with her. He referred to this as *meter o homem na garrafa* (putting a man into a bottle). To break with evil powers that keep persons entangled in depressing relationships of jealousy and hate, the participants had to walk through a bath of salt water in bare feet to neutralize any evil powers. And to strengthen the emotional and bodily experiences of transformation, dramatic sound effects accompanied them as they walked.

Sermons often deal with the meaning of marriage, which is regarded as a relationship between two individuals who serve God. The pastors highlight how, through marriage, a couple are starting a nuclear family by leaving their extended families to live their own lives. Practical tools are given: the couple should live as far away from their kin as possible and, contrary to local custom, the wife should not follow the advice of her mother-in-law but that of her husband. A husband should take care of his wife and make her happy and the two should spend time together, go to the cinema or on holiday. Women are advised to note the culinary tips in *Folha Universal* as well. Pastors criticize local marriages for the dependency they create between the couple and the extended family that, in their view, hinders the establishment of a healthy Christian family. In addition, pastors underscore the importance of the role of sexuality in marriage and stress its pleasures. Sexuality in marriage not only has reproductive goals but should be something both the man and the woman can enjoy. Women are encouraged to play an active role during sex. In the teachings, the presence of love in a relationship is promoted and love means that both parties should try to understand each other and make each other happy. The pastors give examples from their own lives in Brazil.[17] A pastor started a therapy session devoted to trust by saying that Mozambican men are very shy. "You have to step up to a woman when you like her. Make contact and talk to each other."[18] To sum up, one important component of the various discourses and practices at the therapy is to confront local forms of

Figure 10.1 "Therapy of Love" being broadcast on the Universal Church's Miramar TV channel; © Rufus de Vries.

Figure 10.2   The bay of Maputo is a famous spot for wedding photos; © Rufus de Vries.

love as a way of disconnecting from it. To this end, the foreign, transnational connection of the Brazilian pastors appears to be crucial.

Transnational Pentecostal Connections

Several scholars have pointed to the emergence of transnational spaces that are not necessarily shaped by international migration but by processes of communication and exchange generated by capital expansion and the Internet or other forms of communication between specific nation-states (Clifford 1992; Appadurai 1996; Hannerz 1996; Meyer & Moors 2006). These studies analyze how citizens can develop identities that are not necessarily national, for example, through the development of subjectivities and identities based on ideas, customs, practices, and emotions that come to nation-states via travelers, television, and the Internet. It is this idea of people becoming transnational by engaging in mobile structures, cultures, and ideologies that is relevant in the case of the Brazilian Pentecostal churches in Mozambique. Mozambican converts continue to live in their own society while participating in a setting where relations are being developed and maintained that link Brazilian and Mozambican societies by a Pentecostal bridge. This transnational link offers possibilities for identification and loyalty that allow Mozambican Pentecostals to bypass or confront national and/or local domains (Rudolph & Piscatori 1997; Beyer 2001; Corten & Marshall-Fratani 2001; van Dijk 2006).

This process takes on special dimensions in Brazilian Pentecostalism in Mozambique in that converts' cultural nearness to the local society appears critical (van de Kamp & van Dijk 2010). Being part of the local society, unlike in situations of migration, many converts struggle to understand how their Pentecostal morality and spirituality can remain unaffected by local circumstances, powers, or cultural realities. They want to become independent of locally binding forces, that is, to be more culturally and socioeconomically mobile and to cross boundaries. The counseling sessions, particularly the therapy of love, are a key example of the Pentecostal way of confronting local practices through transnational connections. As outsiders, Brazilian pastors force converts to transcend local connections, linking their spiritual power to break through the "silence" surrounding sexuality and love. If people want to develop an independent attitude toward their kin, they have to overcome not only cultural but also spiritual boundaries. The process of therapeutic counseling thus involves the transcendence of spiritual boundaries, for instance, by speaking up about witchcraft and exorcizing spirits. A transnational linkage enhances the aspect of breaking open national and local fields that is an intrinsic aspect of Pentecostalism in Africa (Meyer 1998; van Dijk 1998, 2006).

In this framework, the therapy of love involves disconnecting techniques: openness on intimate matters discloses a contestation of local ideas and practices of love, sexuality, and marriage as a necessary condition for change. People have to learn to think and speak openly about sexuality and relationships. The crucial role of kin in marriage is criticized, and jealousy, distrust, and divorce are seen to result from witchcraft. Abuse is not disclosed in a family setting but on the platform at church. Couples are made to apologize to each other, are told not to involve kin in the settling of disputes, and money to prepare for marriage is not exchanged with kin but presented in church (van de Kamp 2010). At the same time, this transnational Pentecostal space offers converts a place to reflect, talk, and develop new attitudes. Marta (a 23-year-old university student) was impressed with remarks in a pastor's sermon on how a husband and wife should look for a place to live that is far away from kin. She compared it with my situation. My husband and I were living in Mozambique without our families and she concluded that we were learning to be independent and responsible for our own lives. She immediately discussed the issue with her boyfriend as they were in the process of buying land on which to build a house. Pentecostal women find the teachings about husband-wife relations instructive, especially regarding personal development. Lila (a 32-year-old saleswoman) told me that she learned that "we are not slaves [*escravas*] like women are according to Mozambican tradition. I have learned to be independent. It is good, for women, to study and work." Even though a woman's caring role in the family and obedience to her husband are underscored, a woman's individual success is a crucial element of her conversion, as is the transgression of local gender categories (cf. Soothill 2007: 35ff).

The most attractive and exciting aspects of the transnational Pentecostal domain remain the practice of openly learning, and showing and realizing love. When talking about these aspects, women giggle, laugh, and blush. How do they engage with the specific bodily aspects of the therapy? How do they incorporate the transnational connecting techniques of love?

Disciplining the Body in a Sensational Chain of Love

What is striking about women's participation in the *terapia* is how the bodily practices they perform affect their love lives. It would appear that open demonstrations of love generate forces that make love happen. Love comes into existence by using one's body and senses. For example, while telling me about the previous evening's therapy,[19] Célia (a 27-year-old receptionist) said unbelievingly: "I touched a wedding dress [on the platform in the church, LvdK] and am now *determined* to get married" [emphasis LvdK].[20] She explicitly participated in the realization of romantic love by touching the dress and, as such, attached herself to the project of love, which would materialize in a happy marriage. She assured me—and herself—that she was "living in her victory," a popular Pentecostal saying, and that the ideal husband was waiting for her: she only needed to finish her studies and participate in the therapy. I soon learned not to ask about the realization of such plans. It is faith in miracles happening through the enactment of particular therapeutic techniques that confirms that it will occur. My questions were interpreted as a sign of disbelief and I should also walk in a victorious mood, converts told me. Bodily involvement in the therapy is a promise and evidence that love is in the air. However, one has to train one's body and become sensitive to a new culture of love that develops by participating in a chain of connected weekly therapies (*corrente*).

Every convert needs to follow a chain of connected therapeutic techniques to effectuate a new relationship. Every month, or once every six to eight weeks, the pastors choose a specific theme to focus on, such as finding a husband or true love. At the start of the chain, special "therapy-of-love envelopes" are distributed that are linked to a pledge (*propósito*). A new envelope can be filled with money for a specific goal for each *corrente*, for example, finding a *namorado*, and four or more connected therapies have to be followed. In addition, the pastor sets up a particular programme of fasting, praying, and tithing with participants to fight for success in love. Converts have to pray at specific times, try to talk to a potential partner, and keep the envelope under their pillow. The *corrente* ends with a special financial donation that has to be presented at the church's altar on the date specified.

A person's dedication and sincerity have to be proved and the specific requirements of this "treatment" need to be carefully executed. During the chain of therapies, the articulation of love depends on the performance of different formats, for example, putting one's hands on one's heart, walking under a heart, showing oneself to a possible partner, or embracing and kissing. By doing so, converts can be filled by the Holy Spirit who will make them feel and realize love. Recent anthropological work on religion, the body, and the senses has focused on the explicit strategies through which a religious "habitus" is acquired (Hirschkind 2001; Mahmood 2001, 2005; de Witte 2008; Meyer 2009; Klaver 2011). It is by instigating bodily and sensory disciplines that particular religious feelings and responses are raised. For example, the use of music during the therapy and services mediates the personal experience of the Holy Spirit. The physical quality of sound, such as the beat, helps one to feel the Spirit (Marks 1999 in de Witte 2008: 141).

The body itself manifests the presence of the Holy Spirit—or evil powers—when people fall over, scream, cry, and speak in tongues. Particular bodily formats mark spiritual power and one needs to learn how to experience spiritual power and adjust to it. In the framework of the therapy, the power of the Holy Spirit becomes visible in a loving body. The capacity to create a romantic lifestyle depends on inhabiting certain Pentecostal techniques, such as a "new heart." Célia explained that she needed to change the attitude of her heart otherwise she would not succeed in marrying: "God is preparing a husband for me. One day I will marry. My heart is my problem, the devil is there, I need to take care, I need to combat my old way of doing." Sometimes she sent a nasty message to her boyfriend out of the blue: "I really need to control this kind of behaviour. Today, the pastor asked who wanted a new heart. I went forward." She does things she does not want to but by going forward, walking under a heart, or touching a heart, she receives and creates a new heart that will change her behavior.

If one fails to marry or have a happy marriage, it marks one's inability to inhabit the techniques and to form a converted body. One has to rectify the situation by entering a new cycle of performances, a new *corrente* of therapies (cf. Mahmood 2005: 161–167). In other words, it is not only the sign of the Holy Spirit's presence in one's bodily performance that develops into love; an entire manner of being and acting is also required (cf. Campos & Gusmão

2008). By following the chain of bodily formats, love becomes authorized. When single (wo)men stood on the platform and had to seek the Holy Spirit with a hand on their hearts, a particular gaze, happening, or revelation could be an indication from the Spirit that they stood face to face with their future spouse. Converts told me how pastors pointed out who they should marry, which particularly happened with *obreiros/as* (pastors' [fe]male assistants). There were various cases of *obreiras* who soon married an *obreiro* who then became a pastor. They hardly knew each other but since the Holy Spirit showed them that they would be husband and wife and they kissed each other before the altar, true love was guaranteed.

Taking all this together, the Pentecostal bodily formats to love demonstrate a particular relationship between disciplinary structures and agency. It seems that the agency of the converts is involved in how they embody authorized Pentecostal forms. Foucault's analysis of subject formation is useful in understanding the embodiment of divine power. His 1998 [1976] work highlights disciplining and controlling structures in society, such as his study of sexuality. According to Foucault's (1988) elaborations of "technologies of the self," disciplining power cannot be understood solely on the model of domination with a singular intentionality, for example, of the pastors, that is enforced upon others, namely, the converts. Technologies of the self are forms of knowledge, strategies and techniques that "permit individuals to effect by their own means or with the help of others a certain number of operations on their own bodies and souls, thoughts, conduct, and way of being, so as to transform themselves" (*Ibid.*: 18). In other words, disciplining powers, such as therapeutic techniques, allow agents to define their self, in this case in relation to love. Love is produced by the connections the convert is involved in that simultaneously facilitate him/her to create love. However, the individual's responsibility to enact love also means that if he/she does not perform the techniques well, and thus is not able to realize the love propagated by Brazilian Pentecostalism, he/she becomes disconnected from this particular form of social change.

The Ambiguities of the Pentecostal (Dis)connections of Love

The majority of the participants in the therapy are single women. They are hoping to find a Pentecostal partner because he would be trained to be faithful and caring toward his wife. Since there are more women at church than men, women complain that there are insufficient men to engage with. Consequently, their involvement in the therapy does not often result in a loving relationship. Even though being married is not the only ideal of women who attend the church as professional success is equally important, the underlying idea behind most counseling sessions is that being married is better than being single and, for most women, marriage is the ultimate ideal. However, this is an area that is far less controllable than education and work. Engagement with transnational Pentecostalism is thus even more crucial as finding a perfect partner requires a miracle of sorts (cf. Frahm-Arp 2010).

The result of involvement in the therapy is that women feel uncomfortable if they remain single. When they speak in terms of "God gives me the victory" or "God gives me a miracle," they are implicitly saying they would like to have a partner. However, when we became closer friends, it was obvious that the expected boyfriend had never materialized. Célia spoke about her boyfriend but was vague about their current relationship and he seemed to be a fictitious character. The only thing she said was that they did not have much time to meet as she needed to study and was working full time. She was sure that God was preparing her marriage, so in that sense she was victorious. Everything depended on the converts' performance of faith, thus of their embodiment of connecting techniques and whether the constraints of a single life or a non-Pentecostal marriage evolved into a fruitful element in the process of producing love. Several converts gave the impression of being able to master these dilemmas in their reflections on the therapeutic sessions as a way of learning from difficulties and mistakes.

However, I also met converts whose position had become precarious. When Laura married a pastor she hardly knew, her father refused to attend the wedding. She felt increasingly lonely because contact with her family decreased and, at the age of 19, she abandoned her studies because she had to work for the church as the pastor's wife. Suddenly her whole life was being lived in church and when she wanted to go out, she had to ask permission. She had not realized beforehand how it would change.[21] A Mozambican pastor at the Universal Church told me how he had been selected to become an assistant pastor to a Brazilian pastor on condition that he was married.[22] Within a few months, he had married the sister of a good friend of his who was also an *obreira*. The marriage did not work out but he could not be separated from her if he wanted to continue working as a pastor. He commented: "When we marry, we think that the relationship will become good [i.e., love will materialize after taking the right steps LvdK], that we will grow towards each other. I married too hastily, it was a wrong decision."

Preoccupation with the activities of either the Holy Spirit or evil spirits in converts' lives can further disrupt marriages and relationships. According to the same pastor, his wife had a demon. When she started beating him in church, the other pastors also came to the conclusion that she had a bad spirit in her. The tensions in their relationship intensified when the wife's attitude did not change. Converts experience conflict between their attempts to inhabit the prescribed formats of affection and romance, and the control of their behavior by spiritual forces. Demons can imbue the body in much the same way as the Holy Spirit can. Despite the emphasis on people's responsibility to work on the transformation of their body, pastors also stress that converts cannot blame their partners because the Devil is their adversary (cf. Soothill 2007: 209–218). Jealousy, domestic violence, and infidelity are caused by evil spirits. To ensure a successful marriage, women are thus encouraged to become "violent": they have to be God's soldiers and confront the Devil regarding his control over their husband's behavior (van de Kamp 2011b).

As a result, some women, like the pastor's wife mentioned above, enter into quite an ambivalent relationship with their husbands, who they see as being imbued with an evil spirit, and find it difficult to be intimate with their partners. Pastors normally advised their members not to talk about demons with partners who were not Pentecostal because it would worsen the situation. Instead, born-again women take responsibility for the situation and use spiritual powers to convert their partners. As they pray and perform the love techniques, they demonstrate how they are able to appropriate new life forms, which will then materialize in their partners.

To summarize, the creation of specific sensory and bodily perceptions in the therapy of love means that converts and pastors downplay other senses or "anaesthetize" them (Verrips 2006; cf. van Dijk 2009). The transformation of converts' sensory modes leads them to cut their emotional bonds. To realize love, converts demonize their partners and silence any problems. The Pentecostal techniques of confrontation that are part of the public display and realization of love impede the sensitivity to (religious) forms of dialogue and consensus. By doing so, they remain out of touch with the reality of their partners and kin and become far removed from the "romantic love" being propagated by the transnational Pentecostal therapeutic connections of love.

Conclusion

Mozambican youth are actively involved in developing new sexual and marital practices and relations. Urban, educated women in particular are seeking possibilities to openly discuss courtship, marriage, and sexuality and to appropriate novel ways of loving and relating that are compatible with their new socioeconomic position. In a society where affection is generally expressed in hidden ways, viewing images of people showing explicit affection in Brazilian Pentecostal churches is making a powerful impression on them. By disciplining themselves through the explicit bodily forms of the *terapia do amor,* such as kissing and embracing in public, converts aim to create romance, happiness, and fidelity through the power of the Holy Spirit.

Enacting specific bodily and sensory techniques by participating in a chain of therapies increases one's feelings toward potential partners and effectuates marriages. This shows how transnational religious connections are being embodied to effectuate a revolutionary change. The transnational nature of the connection is used to create distance toward the shortcomings of local connections of love, sexuality, and marriage, and to shape a new field of romantic love.

The Brazilian Pentecostal therapy of love shows connection as a strong disciplining force. By appropriating connecting techniques, converts have to show that they are making progress. They have to confront their families openly, amongst other things by denouncing witchcraft practices on the church platform and then they must remain connected to the transnational field by following chains of therapies. The newly appropriated techniques should materialize in romantic relationships and successful marriages, and women who actively engage in the therapy are able to open up a new domain in which they can critique local customs, reflect on alternative modes of love, and realize a desired relationship and marriage even at the cost of distancing themselves from their families and locally accepted behavior. To others, however, the realization of love through the performance of Pentecostal techniques is frustrated by demons that may still control the bodies of converts and those of their partner and kin. The subsequent spiritual battle clearly affects relationships and the struggle against the devil downplays sensitivity to alternative formats of creating love. Living with demonized partners creates relational distance, not passion. The therapy of love as a connecting device can therefore also disconnect people from desired social change.

Notes

1. The most recent development in Christianity is the emergence of Pentecostal and Charismatic churches and movements. There is a variety in the kinds of Pentecostal churches (for Africa, see Kalu 2008 and Meyer 2004). This chapter focuses on the Brazilian neo-Pentecostal churches that are currently active in Mozambique.

2. There is a much larger literature on therapeutic traditions in Africa. See, for example, the special issue on the histories of healing (Schumaker *et al.* 2007).

3. The Brazilian Universal Church arrived in Mozambique in 1992 at the end of the country's 16-year civil war when the socialist period was drawing to a close and neo-liberal economic structures were being introduced (Cruz e Silva 2003; Freston 2005). Other Brazilian Pentecostal churches also arrived, such as the Pentecostal church called God is Love. Another popular one was Maná, coming from Portugal but often considered to be Brazilian. Brazilian missionaries also worked in some local churches, such as in the Pentecostal Assemblies of God. See van de Kamp (2011a).

4. Services have a special focus every day—Monday: finance and business; Tuesday: health; Wednesday and Sunday: conversion/personal growth by means of the Holy Communion; Thursday: the family; Friday: the exorcism of demons/spiritual war; Saturday: the therapy of love.

5. When "the therapy" is mentioned in this chapter, I refer to the *terapia do amor*.

6. With slight variations, this "kiss-and-embrace" moment happened almost every time I visited the therapy in 2006 and 2007, and again in July 2008.

7. Twenty-five months of ethnographic fieldwork in Mozambique took place between 2005 and 2008 as part of doctoral research that was sponsored by the Netherlands Organisation for Scientific Research (NWO).

8. With the increased Brazilian presence in Mozambique and the liberalization of the mass media since the 1990s, the Brazilian prominence in the Mozambican media has grown (Power 2004: 278), notably, the diffusion and popularity of *telenovelas*. In conversations about romance, love, and marriage, the Brazilian *telenovelas* and the Brazilian churches were important sources of comparison and inspiration. There is no space here to deal with the intertwining of Brazilian *telenovelas* and Pentecostalism. See Oosterbaan (2006) and van de Kamp (2011a) for more on this subject.

9. This depends on the stage of the *namoro*. Usually the *namorado* will present his *namorada* to his kin sooner. The boyfriend can only be received at the girlfriend's home after the relationship has become official following a particular ceremony, an initial step in the *lobolo* (brideprice) procedure, that marks the serious nature of the relationship (Granjo 2005; Bagnol 2006).

10. This is also related to the bottles *curandeiros* (local healers) use for their magical substances. Bottles are known to hold energetic powers, like spiritual forces, and women in particular visited *curandeiros* to get medicines that would bring a man into their power. But often this man already had a *namorada* or *pitas*.

11. See Cruz e Silva *et al.* (2007) for the influence of processes of modernization on the current role (or the absence) of rites of passage in relation to sexuality in various parts of Mozambique.

12. The names I use are not the converts' real names.

13. Interview, June 28, 2006.

14. For example: "Is It Possible to Stay Happily Married Forever?" (*Folha Universal*, February 13–19, 2005: 6) and "Love and Passion" (Macedo 2005: 63). See also http://terapiadoamor.org/

15. July 26, 2008.

16. January 28, 2006.

17. The pastors were mostly Brazilian and their assistants were Mozambican. I have also seen a pastor from Angola who led the therapy.

18. July 26, 2008.

19. Conversation held on June 24, 2007.

20. During therapies and services participants often had to say "I determine that this and that will happen" in order to make it seem real. Neo-Pentecostal churches

follow the Faith Theology by attributing spiritual power to the spoken word. The idea of the power of the word is known from the Word of Faith Movement that originated in the United States under the leadership of founding fathers Kenneth Hagin and Kenneth and Gloria Copeland (Anderson 2004: 39–165).

21. Interview with Laura's sister, February 7, 2007.
22. Interview with Laura's sister, February 7, 2007.

References

Anderson, A. (2004) *An Introduction to Pentecostalism: Global Charismatic Christianity.* Cambridge: Cambridge University Press.

Appadurai, A. (1996) *Modernity at Large. Cultural Dimensions of Globalization.* Minneapolis: University of Minnesota Press.

Bagnol, B. (2006) "Gender, Self, Multiple Identities, Violence and Magical Interpretations in 'Lovolo' Practices in Southern Mozambique," Ph.D. Thesis, University of Cape Town.

Becker, F. & P. Wenzel Geissler (eds) (2007) "Faith and AIDS in East Africa," *Journal of Religion in Africa* 37(1): 1–149.

Becker, F. & P. Wenzel Geissler (eds) (2009) *Aids and Religious Practice in Africa.* Leiden: Brill.

Beyer, P. (ed.) (2001) *Religion in the Process of Globalization.* Wurzburg: Ergon Verlag.

Bochow, A. (2008) "Valentine's Day in Ghana: Youth, Sex and Fear between the Generations," in: E. Alber, S. van der Geest & S. R. Whyte (eds), *Generations in Africa.* Hamburg: LIT Verlag, pp. 418–429.

Bochow, A. (2010) *Intimität und Sexualität vor der Ehe. Gespräche über Ungesagtes in Kumasi und Endwa, Ghana.* Berlin: LIT Verlag.

Burchardt, M. (2009) "Subjects of Counseling: Religion, HIV/AIDS and the Management of Everyday Life in South Africa," in: F. Becker & P. Wenzel Geissler (eds), *Aids and Religious Practice in Africa.* Leiden: Brill, pp. 333–358.

Burchardt, M. (2011) "Challenging Pentecostal Moralism: Erotic Geographies, Religion and Sexual Practices among Township Youth in Cape Town," *Culture, Health and Sexuality* 13(6): 669–683.

Campos, R. B. C. & E. H. Gusmão (2008) "Celebração da Fé: Rituais de Exorcismo, Esperança e Confiança, na IURD," *Revista AntHropológicas* 19(1): 91–122.

Clifford, J. (1992) "Traveling Cultures," in: L. Grossberg, C. Nelson & P. A. Treichler (eds), *Cultural Studies.* New York: Routledge, pp. 96–116.

Cole, J. (2010) *Sex and Salvation: Imagining the Future in Madagascar.* Chicago: University of Chicago Press.

Collier, J. (1997) *From Duty to Desire: Remaking Families in a Spanish Village.* Princeton: Princeton University Press.

Comaroff, J. & J. Comaroff (1991) *Of Revelation and Revolution: Christianity, Colonialism, and Consciousness in South Africa. Vol. 1.* Chicago: University of Chicago Press.

Comaroff, J. & J. Comaroff (1997) *Of Revelation and Revolution: Christianity, Colonialism, and Consciousness in South Africa. Vol. 2: The Dialectics of Modernity on a South African Frontier.* Chicago: University of Chicago Press.

Corten, A. & R. Marshall-Fratani (eds) (2001) *Between Babel and Pentecost. Transnational Pentecostalism in Africa and Latin America.* London: Hurst & Co.

Cruz e Silva, T. (2003) "Mozambique," in: A. Corten, J-P. Dozon & A. Pedro Oro (eds), *Les Nouveaux Conquérants de la Foi. L'Église Universelle du Royaume de Dieu (Brésil).* Paris: Karthala, pp. 109–117. Published in Portuguese by Paulinas, São Paolo.

Cruz e Silva, T., X. Andrade, M. J. Arthur & C. Osório (2007) "Representations and Practices of Sexuality among the Youth, and the Feminization of AIDS in Mozambique," Research Report, Maputo: WLSA Mozambique.

de Witte, M. (2008) "Spirit Media. Charismatics, Traditionalists, and Mediation Practices in Ghana," Ph.D. Thesis, University of Amsterdam.

Etnofoor (2006) "Romantic Love," *Etnofoor* 19(1).

Feliciano, J. F. (1998) *Antropologia Económica dos Thonga do Sul de Moçambique.* Maputo: Arquivo Histórico de Moçambique.

Foucault, M. (1988) "Technologies of the Self," in: L. H. Martin, H. Gutman & P. H. Hutton (eds), *Technologies of the Self.* Amherst: University of Massachusetts Press.

Foucault, M. (1998) [1976] *The History of Sexuality. The Will to Knowledge, Vol. 1.* UK: Penguin.

Frahm-Arp, M. (2010) *Professional Women in South African Pentecostal Charismatic Churches.* Leiden: Brill.

Freston, P. (2005) "The Universal Church of the Kingdom of God: A Brazilian Church Finds Success in Southern Africa," *Journal of Religion in Africa* 35(1): 33–65.

Geissler, P. W. & R. J. Prince (2007a) " 'Life Seen': Touch and Vision in the Making of Sex in Western Kenya," *Journal of Eastern African Studies* 1(1): 123–149.

Geissler, P. W. & R. J. Prince (2007b) "Christianity, Tradition, AIDS and Pornography: Knowing Sex in Western Kenya," in: R. Littlewood (ed.), *Knowing and Not Knowing in Anthropology.* London: UCL Press, pp. 87–116.

Giddens, A. (1992) *The Transformation of Intimacy. Sexuality, Love and Eroticism in Modern Societies.* Oxford/Cambridge: Polity Press.

Granjo, P. (2005) *Lobolo em Maputo: Um Velho Idioma para Novas Vivências Conjugais.* Porto: Campo das Letras.

Gune, E. (2008) "Momentos Liminares: Dinâmica e Significados no Uso do Preservativo," *Análise Social* XLIII(2): 297–318.

Hannerz, U. (1996) *Transnational Connections. Culture, People, Places.* London: Routledge.

Hawkins, K., F. Mussá & S. Abuxahama (2005) *"Milking the Cow": Young Women's Constructions of Identity, Gender, Power and Risk in Transactional and Cross-Generational Sexual Relationships.* Mozambique: Options Consultancy Services & Population Services International (PSI).

Hirsch, J. S. & H. Wardlow (eds) (2006) *Modern Loves: The Anthropology of Romantic Courtship & Companionate Marriage.* Ann Arbor: University of Michigan Press.

Hirschkind, C. (2001) "The Ethics of Listening: Cassette-Sermon Auditioning in Contemporary Egypt," *American Ethnologist* 28(3): 623–649.

Igreja, V. & B. Dias-Lambranca (2009) "The Thursdays as They Live: Christian Religious Transformation and Gender Relations in Postwar Gorongosa, Central Mozambique," *Journal of Religion in Africa* 39(3): 262–294.

Illouz, E. (1997) *Consuming the Romantic Utopia. Love and the Cultural Contradictions of Capitalism.* Berkeley: University of California Press.

Kalu, O. (2008) *African Pentecostalism: An Introduction.* Oxford: Oxford University Press.

Karlyn, A. S. (2004) "Sexual Identity, Risk Perceptions and AIDS Prevention Scripts among Young People in Mozambique," Ph.D. Thesis, University of London.

Karlyn, A. S. (2005) "Intimacy Revealed: Sexual Experimentation and the Construction of Risk among Young People in Mozambique," *Culture, Health & Sexuality* 7(3): 279–292.

Klaver, M. (2011) *"This Is My Desire". A Semiotic Perspective on Conversion in an Evangelical Seeker Church and a Pentecostal Church in the Netherlands.* Amsterdam: Amsterdam University Press.

Lindholm, C. (2006) "Romantic Love and Anthropology," *Etnofoor* 21(1): 5–21.

Loforte, A. M. (2003) *Género e Poder entre os Tsonga de Moçambique.* Lisbon: Ela por Ela.

Macedo, E. (2005) *O Perfil da Mulher de Deus.* Rio de Janeiro: Editora Gráfica Universal Ltda.

Mahmood, S. (2001) "Rehearsed Spontaneity and the Conventionality of Ritual: Disciplines of Salāt," *American Ethnologist* 28(4): 827–853.

Mahmood, S. (2005) *Politics of Piety: The Islamic Revival and the Feminist Subject.* Princeton: Princeton University Press.

Manuel, S. (2008) *Love and Desire: Concepts, Narratives and Practices of Sex amongst Youths in Maputo City.* Dakar: CODESRIA.

Manuel, S. (2011) *Changing notions of sexuality in Mozambique.* Ph.D. Thesis, London: SOAS.

Meyer, B. (1998) " 'Make a Complete Break with the Past': Memory and Postcolonial Modernity in Ghanaian Pentecostal Discourse," *Journal of Religion in Africa* 28(3): 316–349.

Meyer, B. (2004) "Christianity in Africa: From African Independent to Pentecostal-Charismatic Churches," *Annual Review of Anthropology* 33: 447–474.

Meyer, B. (ed.) (2009) *Aesthetic Formations: Media, Religion and the Senses in the Making of Communities.* New York: Palgrave.

Meyer, B. & A. Moors (eds) (2006) *Religion, Media and the Public Sphere.* Bloomington: Indiana University Press.

Nguyen, V. K. (2009) "Therapeutic Evangelism: Confessional Technologies, Antiretrovirals and Biospiritual Transformation in the Fight against AIDS in West

Africa," in: F. Becker & P. W. Geissler (eds), *Aids and Religious Practice in Africa.* Leiden: Brill, pp. 359–378.

Oosterbaan, M. (2006) "Divine Mediation. Pentecostalism, Politics and Mass Media in a Favela in Rio de Janeiro," Ph.D. Thesis, University of Amsterdam.

Osório, M. de Conceição (1997) *Violencia Contra a Jovem e Construção da Identidade Feminina. Relatório de Investigação.* Maputo: Muleide.

Osório, M. de Conceição & T. Cruz e Silva (with collaboration by V. Monjane) (2008) *Buscando Sentidos: Género e Sexualidade entre Jovens Estudantes do Ensino Secundário, Moçambique.* Maputo: WLSA Moçambique.

Paulo, M. (2005) *Jovens, Sexualidade e Educação Sexual no Bairro de Mafalala, Cidade de Maputo, Moçambique.* Maputo: Imprensa Universitária.

Penvenne, J. M. (1997) "Seeking the Factory for Women. Mozambican Urbanization in the Late Colonial Era," *Journal of Urban History* 23(3): 342–379.

Power, M. (2004) "Post-Colonial Cinema and the Reconfiguration of Moçambicanidade." *Lusotopie* 11(1/2): 261–278.

Prince, R., P. Denis & R. van Dijk (eds) (2009) "Christianity and HIV/AIDS in East and Southern Africa," Special Issue, *Africa Today* 56(1).

Ranger, T. (1992) "Godly Medicine: The Ambiguities of Medical Mission in Southeastern Tanzania," in: S. Feierman & J. M. Janzen (eds), *The Social Basis of Health and Healing in Africa.* Berkeley: University of California Press, pp. 256–282.

Rudolph, S. H. & J. Piscatori (eds) (1997) *Transnational Religion and Fading States.* Boulder: Westview Press.

Santos, B. & M. J. Arthur (1994) "Enquanto os Homens Tiverem o Poder Sexual... O Comportamento Sexual e a Expansão do SIDA/DTS na Cidade de Maputo," in: UNESCO, *Eu Mulher em Moçambique.* Maputo: UNESCO & AEMO, pp. 69–81.

Schumaker, L., D. Jeater & T. Luedke (2007) "Histories of Healing: Past and Present Medical Practices in Africa and the Diaspora," Special Issue. *Journal of Southern African Studies* 33(4).

Soothill, E. (2007) *Gender, Social Change and Spiritual Power: Charismatic Christianity in Ghana.* Leiden: Brill.

Spronk, R. (2002) "Looking at Love. Hollywood Romance and Shifting Notions of Gender and Relating in Nairobi," *Etnofoor* 15: 229–240.

Spronk, R. (2006) "Ambiguous Pleasures. Sexuality and New Self-Definitions in Nairobi," Ph.D. Thesis, University of Amsterdam.

Thornton, R. J. (2009) *Unimagined Community: Sex, Networks, and AIDS in Uganda and South Africa.* Berkeley: University of California Press.

UNAIDS/WHO. (2008) *Epidemiological Fact Sheet on HIV and AIDS: Core Data on Epidemiology and Response, Mozambique 2008 Update.* Joint United Nations Programme on HIV/AIDS & World Health Organization.

van de Kamp, L. (2010) "Burying Life: Pentecostal Religion and Development in Urban Mozambique," in: B. Bompani & M. Frahm-Arp (eds), *Development and Politics from Below: Exploring Religious Spaces in the African State.* New York: Palgrave MacMillan, pp. 152–168.

van de Kamp, L. (2011a) "Violent Conversion: Brazilian Pentecostalism and the Urban Pioneering of Women in Mozambique," Ph.D. Thesis, Vrije Universiteit Amsterdam.

van de Kamp, L. (2011b) "Converting the Spirit Spouse: The Violent Transformation of the Pentecostal Female Body in Post-War Urban Mozambique," *Ethnos* 74(4): 510–533.

van de Kamp, L. & R. van Dijk (2010) "Pentecostals Moving South-South: Brazilian and Ghanaian Transnationalism in Southern Africa," in: A. Adogame & J. Spickard (eds), *Religion Crossing Boundaries: Transnational Dynamics in Africa and the New African Diasporic Religions*. Leiden: Brill, pp. 123–142.

van Dijk, R. (1998) "Pentecostalism, Cultural Memory and the State: Contested Representations of Time in Postcolonial Malawi," in: R. Werbner (ed.), *Memory and the Postcolony. African Anthropology and the Critique of Power*. London/New York: Zed Books, pp. 155–181.

van Dijk, R. (2006) "Transnational Images of Pentecostal Healing: Comparative Examples from Malawi and Botswana," in: T. J. Luedke & H. G. West (eds), *Borders and Healers: Brokering Therapeutic Resources in Southeast Africa*. Bloomington: Indiana University Press, pp. 101–124.

van Dijk, R. (2009) "Gloves in Times of Aids: Pentecostalism, Hair and Social Distancing in Botswana," in: F. Becker & P. Wenzel Geissler (eds), *Aids and Religious Practice in Africa*. Leiden: Brill, pp. 283–308.

van Dijk, R. (2010) "Marriage, Commodification and the Romantic Ethic in Botswana," in: M. Dekker & R. van Dijk (eds), *Markets of Well-Being: Navigating Health and Healing in Africa*. Leiden: Brill, pp. 282–305.

Verrips, J. (2006) "Aisthesis and An-Aesthesia," *Ethnologia Europea* 35(1/2): 27–33.

CHAPTER 11

Ajala Travel: Mobility and Connections as Forms of Social Capital in Nigerian Society

Oka Obono and Koblowe Obono

Introduction

Nigerian society is a complex amalgam of systems of stratification and inequality although the question of how far social mobility has changed the norms of its sedentary moral economy has not received the kind of scholarly attention it deserves. This chapter advances the argument that, while human mobility is indeed an integral part of the country's ancestral migrant meta-narratives, with most of its settlements coming into being due to the actions of migrants, mobility has assumed a different form since independence in 1960.

This new pattern of emigration reconnected the country to a wider global society that had previously dominated relations with it in less transactional or equitable terms. Its distinguishing feature was that its self-consciousness was newly styled in a globalizing world. Within this autonomy and in its emergent aberrant forms, it was sometimes undertaken against the legislative will of host countries. The event was encapsulated in a worldview that assessed mobility and the establishment of ties as essential to the formation of social capital and the enhancement of one's material, economic, cultural, and political quality of life.

Modern mobility in Nigeria acquired its acquisitive tendencies within this latter subphase of emigration. Mobility on the scale described, that is, in its transnational character, was perceived as a strategic means for improving welfare during the course of a person's life and it was in general restricted to the emerging elite. For this reason, and in this form, modern Nigerian emigration constituted a break from two major historical patterns of trans-Saharan and trans-Atlantic African migration in that both were forced. The first pattern was driven by business interests in human trafficking and slavery, and the second was short-lived by comparison, taking the form of military conscription of Black platoons to assist Allied efforts during the First and Second World Wars.

In neither of these previous patterns of slavery or conscription did emigration or mobility stem from the autonomous decisions of the migrants, nor was the movement expected by any of the parties to provide direct economic benefits or viable political connections to them. The migrants were, in both patterns, either chattels or conscripts and had no motives or aspirations of their own. They were the victims of forced migration motivated purely and simply by European business and political interests.

The third pattern of historical migration can be distinguished by its self-consciousness. Migratory motivations and aspirations were autonomous in the sense that emigration was self-directed and, possibly, self-initiated. This thus accounts for the organization of the historiography around Ajala. While migration remained a form of collective activity, in its motivations, its autonomy became more sharply defined than what took place earlier as international travel. This is the attribute that has characterized Nigerian emigration since the 1960s and has made it distinctive. Nonetheless, it is true that "until the early 1980s, few African professionals, especially Nigerians, saw emigration as a rewarding option" (Adepoju 2010: 13). But it is equally clear that a critical mass of migrants had begun to flow outward more than 20 years earlier.

The emigration described by demographer Aderanti Adepoju, doyen of African migration studies, was in fact stimulated by worsening economic conditions that were associated with the structural adjustment policies of the military junta in the 1980s. This subphase was an intensification of a process that had already begun and was not its beginning. In this sense, early postindependence emigration was the third main phase of Nigerian international migration. It emanated in that period from strategic thinking reflecting the aspirations of the indigenous elite as they scrimmaged and maneuvered for opportunities in what can be characterized as a postcolonial scramble for Nigeria.

A recent Brazilian article by Maia (2011: 393) argues that "there has been a rich discussion concerning the consequences of the spatial turn for social theory, but both its full potential and the contribution of non-European forms of spatial imagination remain unexplored." This suggests the need for a tradition of geographical thinking that provides different frameworks for global spatial imagination in which:

> The aim is to argue that this perspective helps to de-centre social theory by providing new spatial images that diverge from those related both to the language of the city and the Eurocentric perspective that still characterizes the spatial turn.
>
> (*Ibid.*: 392)

In line with this renewed emphasis on "the power of imagination in geographical discourse" (*Ibid.*: 401), the metaphorical analysis presented in the next section shows that the coincidence of this third wave or pattern of mobility with independence was not unusual. Instead, it was consistent with the emergence of a nonsedentary worldview, with a class that endorsed it. This metaphorical analysis conduces with conventions in linguistic research and communication theory that hold that historical fables and metaphors are germane to comprehending such contemporary forces and phenomena as modern elite mobility in Nigerian society. It resituates individual behavior within a fusion of time and space in which we present some preliminary syntheses of social capital and adaptations of a conceptual framework for studying aspirations proposed by Sherwood (1989). These syntheses help to relate mobility to the discourse of motives, perceptions, and aspirations intended to establish connections and increase social capital in the way in which Bourdieu first formulated the concept.

The Global Mobility of Local Clichés

A hermeneutic assessment of how mobility was redirected and intensified soon after independence when governing class formations were beginning to crystallize shows that this mobility was simultaneously shaped by numerous global factors as well as aspirations that were, to a large extent, endogenous. The assessment relies on stories that need to be decoded before their metaphors of mobility can reveal their connections with the Nigerian state and the global economy. The procedure is based on ideas contained in the article entitled "Meaning through Metaphor: Analogy as Epistemology" by Livingstone & Harrison (1981) that suggested that decoded metaphors were indispensable to the understanding of social behavior. For this reason, this chapter has adopted what is described as the "Ajala" motif as the springboard for its analysis because of the motif's consistency with an approach in social biography.

The metaphorical analysis begins with basic historiography to identify not only reverse migrant flows but also a reverse mercantilism of some deviant African migrants who display the same predatory instinct and mercenary worldview that, in a previous age, made slavery and colonization possible. It is the countercultural perspective that has led to frequent media reports of financial crimes and other misdemeanors of migrants who are so desperate that they seek wealth and/or fame by fraud and subterfuge.

This mobility is a departure from the central trajectory of African emigration, whether in the third wave or the preceding two. In the main, the event is constituted by outward flows of law-abiding migrants who work in host countries to send remittances home while contributing to the maintenance of order and prosperity in their host countries. They regulate systems on both sides, which is why the tendency and process are conceptualized as *the mobility of norms.*

Our metaphorical procedure consists of a cursory review of colloquial expressions that have evolved in Nigeria to describe the imagined exotic character of the returnee or, more generally, the mobile person. We highlight the honor reserved for such returnees and how that honor was comparable to that reserved for respected warriors and eminent hunters at earlier stages in social evolution and community development.

In the sedentary state, where the moral code was strong and norms of propriety critical for societal stability, it was assumed by members of their communities of origin that return migrants had acquitted themselves as worthy moral ambassadors of their home communities in the host country. The historiography shows that wealth amassed through means opprobrious at home (such as theft or prostitution) was not accorded this respect. A range of diffuse negative sanctions were applied to mobility that had such outcomes.

Not only is it possible to trace or periodize the emergence and evolution of social mobility in Nigeria during the third wave, it is also necessary to account for its paradoxical depictions within the contexts of time, space, and stigma. Depending on the scope of mobility—the distance covered or boundaries traversed, whether within local communities or beyond national boundaries—these depictions can be positive or negative. In either case, the paradox is consistent with the assumptions of a sedentary communal philosophy as it encountered change. On the one hand, the expressions denote the excitement of travel and its many material and ideational benefits, while on the other hand, they focus on its less salubrious attributes. The determination of whether one type of movement is good or bad becomes the subject of moral geography as stigma can be realized as a function of space and social distance.

Among terms denoting mobility, the Yoruba *Tokunbo*[1] stands out. With its derivation from the word *okun* (sea), it suggests international travel and, by extension, imported secondhand cars are commonly referred to as *tokunbo*. Its use in this sense is so commonplace that Oladeji (2010) does not problematize it when reporting that:

> commodities that are frequently smuggled in and out of the country include petroleum products, cigarettes, textile materials, and currencies, fairly used cars, "tokunbo" vehicles, fairly used electrical and electronic gadgets, arms and ammunition, rice and groundnut oil. While petroleum is the principal product being smuggled out; the respondents indicated that "tokunbo" vehicles, textile materials, rice, groundnut oil and arms and ammunition are the major goods being smuggled in.

A study of pollution levels in Yoruba-speaking Abeokuta similarly reports that "it is believed (quite erroneously) that the addition of the lead additives is the cheapest way for boosting the octane number of gasoline used by large numbers of old used cars (called Tokunbo locally) imported from the developed countries" (Odukoya, Arowolo & Bamgbose 2000).[2]

In the popular imagination, therefore, human *Tokunbos* had more connections by virtue of the circumstances of their birth, which conferred highly visible economic, cultural, and political advantages on them and, by extension, their families. Whether the word is used to describe men or materials, it generally conjures up the unmistakable image of affluence and well-being. It is a preliminary linguistic means of signifying status and separation, a device for acknowledging progress, inequality, and exclusivity. In this respect, the word *Tokunbo* foregrounds the long association of Nigeria's local communities with the modern discourse of mobility and connections.

Another term that resonates in a similar way is *Ajala Travel*, the leitmotif of this chapter. This was the *nom de guerre* of the legendary Olabisi Ajala who is presented by the media in southwest Nigerian as a hero and exemplar of foreign travel. He was popularized by converted highlife-juju musician Ebenezer Obey (b. 1942) in a song memorializing his exploits as an explorer.[3] It portrays him as traveling "all over the world," presumably on a motorized scooter (Vespa),[4] a swashbuckling Gulliver but with no documented accounts of his adventures. According to one commentary, "his story was like the Nigerian version of Mungo Park, a voyager that navigated some troubled oceans with a boat . . . Ajala's globe-trotting story in the 60s became a lexicon in Nigerian language while allusions are made to people who love travelling as 'Ajala' " (Odeyemi 2004).

Although the story has achieved iconic status in Nigeria, we view the use of a Vespa by Mr. Ajala as problematic. It was a poor and impractical strategic choice for such a mission and we think its real significance lies in the semiotic nuance it introduces into the account. Its engineering was not designed for a mission on such a scale and across rugged territory.

With its practicality thus in question, the Vespa is to be understood metaphorically as the embodying emergent aspirations of a newly mobile and independent nation in the 1960s when Nigeria was just freeing itself of British colonial rule. It was the metaphorical vehicle by which change could come and in relation to which a generation of nationalists pushed themselves to great endeavor, to become mobile even with the most rudimentary methods. The Ajala personified the courage needed to initiate this.

The Vespa was integrated into Ajala narrative as part of this basic allegorical design—whether Ajala was in fact real or not, or undertook travels on the scale he was purported to have done. This is not quite the crux of the matter. In an erstwhile sedentary society, tales of travel within Europe or in societies with an advanced transportation infrastructure (good roads and railway networks) might hold a certain psychic appeal or fascination for a domestic audience. Ajala might merely have moved across a few closely linked territories covered by the modern Schengen Treaty—situated just mere hours from one another—and that would have been sufficient to accord him the status of someone who "travelled all over the world." The distances covered might have been exaggerated by the absence of communication technology, such as the Internet, fax, and/or the mobile phone. This exaggeration might in turn have been intensified by the exotic stories he brought home with him. In any case, it is doubtful whether he literally traveled the world, as Obey claimed.

There are practical doubts about the specifics of Ajala's adventures. It is easier to appreciate the story in the symbolic light of an avant-garde spirit in control of both the narrative itself and the emergence of an indigenous but mobile elite in Nigerian society at that time. It was a *Weltanschauung* that stimulated waves of migration to Europe and America by citizens in search of the proverbial Golden Fleece. These were the citizens who would in due course return to establish the country's first republican bureaucracy and lay the foundations for its operation within a democratic state. The Ajala motif was the personification of this rising postcolonial consciousness while his Vespa was the index of the difficulties associated with realizing it in practical economic, political, and institutional terms. Assessing these metaphors for their meanings and contextualizing epistemology by means of analogical reference (Livingstone & Harrison 1981) is a valid procedure because, in the context of this metaphorical analysis, the symbolism of his travels is more meaningful than the actual facts involved.

Negative Mobility and Sedentary Stigma

The preceding sections discussed how the words *Ajala* and *Tokunbo* both imply movement and stand as status signifiers. They symbolize the presence of something exotic yet indigenous as well as, of productive migrant histories and rich family backgrounds without which neither outcome would have been possible. *Tokunbo*, in particular, foregrounds hybrid identities that confer status on people designated this way or identified by similar appellations in Nigeria's ongoing postcolonial relations with itself.

Along these lines, recent commentary suggests that the entire third trend of African emigration could be reversed. Obono (2010) notes that:

> The European Social Survey, which has documented attitudes on a biennial basis since 2002, confirms that Europeans' views of immigrants have deteriorated drastically and steadily in this decade. Immigrants are seen as posing fierce competition for scarce benefits. Many observers agree that, in light of the new face of immigration policy in the customary destinations of African migrants, an eventual African exodus is inevitable. Already, Nigeria registers a continuous stream of deportees from many countries. Partly in response, many Nigerians abroad are buying up real estate . . . Funky looking estates designed for upwardly mobile families are springing up along expressways everywhere in the country. They are stimulated by forecasts of mass return as adversity grows in receiving countries. If corpses are still brought back to Nigeria for burial, it demonstrates that the return of the living awaits auspicious conditions.

In this light, there is a systemic irony that constitutes local variants of these *ajala* and *tokunbo* movements as something negative. These variants are couched in the admonitory pidgin *waka-about* (walk about) and its corollary *Amebo*. While *Tokunbo* and Ajala are viewed positively in their mobility, the local *waka-about* is seen as a regular menace to society. A rumormonger and meddlesome busybody, he/she is compulsively peripatetic. A restless nosy parker driven by an unaccountable need to pry into people's affairs, he/she is what the Yakurr call an *oseŋa-seŋa* (someone who walks about doing no good).

By spreading rumors, the *oseŋa-seŋa* helps maintain social order and systemic stability and keeps the members of society connected to a false information grid or network. The actions of an *oseŋa-seŋa* thus promote negative solidarity among victims who are united by their mutual adversity occasioned by his/her actions. This happens when the *oseŋa-seŋa* doubles as an *amebo* (the bearer of tales or quintessential gossip) or what is known in Yoruba as an *olofofo*.[5]

Such appellations and terminology depict a negative view of mobility and suggest a spatial dimension to the assignation of value or stigma to different types or patterns of mobility. Movements across national boundaries are heralded as more prestigious than those within domestic or national boundaries. There seems to be an inverse correlation between distance covered by travel and the probability of stigma.

The questions raised by these dynamics are clearly more nuanced and complicated than this but this chapter is only examining these clichéd views of social mobility to derive hypotheses about the spatial dynamics of identity and negation. By converging the constructs of aspiration and social capital theory, social mobility is shown to both cause and effect connections established across space and time by Nigerian migrants. Mobility is thus a predictor of social capital developed in the course of establishing those connections as well as a means of its expression among the elite.

Discussion: Mobility, Aspirations and Social Capital

Nigerian society is in general characterized by ubiquitous norms of inequality. Its historical forms of stratification have ranged from the condemnable caste systems among the Igbo in the southeast to forms of serfdom in the north. Other structural categories include gerontocracy, theocracy, and patriarchy. In all cases, people are segregated into unequal divisions that made status a criterion for accessing resources. There is a built-in systemic impulse to stratify. Hence, in a well-worn anecdote, it is said that if three Nigerians were marooned on an island with no provisions, their first order of business would be to appoint a Chairman—anything to trace, identify, reinforce, and perpetuate lines of inequality.

The result of this at a macrocosmic level is the size of the country's modern government with an executive cabinet of 42 ministers, a bloated civil service, and patron-client networks that fuel much of the official corruption for which the country is, unfortunately, so famous (Smith 2008). This mix of stratification systems would appear to be part of the reason behind the country's difficult transition to proper modern governance and democracy, and it is into this complex of systems that social mobility falls.

The phenomenon is usually measured in terms of occupational prestige or educational attainment but these two attributes are by no means exhaustive or exclusive. In recent models of countercultural mobility, that is, mobility that results from or leads to the suppression of communal values, there may be no clarity as to the migrants' occupational status so it cannot be measured. Being criminal, it might be secret. Identities may be concealed. Moreover, educational attainment might not have been part of the original motive for the movement so it too might not always be a good measure of status change. In other words, while social mobility is usually measured by the variables of education and occupation, its emergent forms may involve neither.

One report notes that "interpreting the narratives of Moroccan migrant women in the Netherlands, alternative definitions of social mobility are discerned that go *beyond* formal schooling or paid work and which contribute to a broader definition of class and 'social upgrading'." It describes cross-generational differences in how migrant women subscribe to dominants definitions of mobility in Rotterdam (van den Berg 2011: 503), important differences that demonstrate that the social contexts of aspirations and mobility may vary widely. As shown elsewhere (Obono 2008; Obono & Obono 2009, 2010), there are important intergenerational differences of perception and practice among mothers and daughters on the same behavioral questions and attitudes that are linked to social mobility.

In instances where sedentary norms and livelihood systems give way to physical and social mobility, there are new contentions with alternate (sometimes criminal) economies as people develop solidarities and maintain networks made possible by new technologies. In such situations, economic prospects are linked to the scope and strength of one's connections and access to social media, which also facilitate criminal activities. The conjuncture between technology and criminality is a crucial phase in Nigerian society that was previously built around fixed, inflexible, and mechanical networks of kinship and other primordial connections and solidarities. These changes have intensified with the emergence of new technologies but they have also created new and increased vulnerabilities.

While mobilities can consolidate connections, they also help produce mass alienation by interrupting harmony and bringing about new possibilities in the future course of human relations. Through access to the information highway, notions of mobility and connectivity have become more fluid and Nigeria is reputed to have the highest teledensity in Africa. There are substantial departures from the physical sense in which connections and mobility were previously understood. The cartographies have been resocialized into a more conscious awareness that appropriates connections, especially in transborder forms, as a requirement for occupational success.

Understanding this tendency to move in order to generate (greater) material and social benefits can be aided by the concept of social capital. Defined as the "resources that emerge from one's social ties" (Portes & Landolt 1996), the popularity of social capital as a concept has been accompanied by "increasing controversy about its actual meaning and effects" (Portes 2000: 1). This is why Fukuyama (2002: 27) considers it "difficult to define." While aggregate applications seem to endorse its use among World Bank economists and academic political scientists, who attribute it to whole state systems or organizations, its original Bourdieuan sense was far more practical and instrumental in that it could be reduced to the level of a tradable and fungible possession. It dealt with benefits that accrued to individuals and their families as they intentionally built their relationships with others around sets of individual and group aspirations and goals (Bourdieu 1985).

Noting that capital can assume any of several mutable forms—cultural, economic, functional, linguistic, personal, local, professional, social or symbolic—Bourdieu (1985: 248) provided the first definition of *social* capital as the "aggregate of the actual or potential resources which are linked to possession of a durable network of more or less institutionalized relationships of mutual acquaintance or recognition."

Nevertheless, in its purest form, the concept is strong because of its instrumental focus on "the benefits accruing to individuals by virtue of participation in groups and on the deliberate construction of sociability for the purpose of creating this resource" (Portes 1998: 3). In other words, "the core idea of social capital is that social networks have value" (Ecclestone & Field 2003: 267). Viewed in this way, the concept facilitates our understanding of how people acquire and apply skills and knowledge, which is part of its broad analytical appeal.

Hence our argument that since "social capital of any significance can seldom be acquired, for example, without the investment of some material resources and the possession of some cultural knowledge, *enabling the individual to establish relations with others*" [emphases added] (Portes 2000: 2), the utility of the concept for social mobility requires a perspective in aspirations theory. Without this perspective, the concept may be "further clouded by the fact that it is not always easy to judge whether or not a given set of relations constitute 'capital,' an investment that may yield a desirable outcome" (Bankston 2004: 178). This is why, in our view, the joint focus of both concepts (social capital and aspiration) on the individual, conscious material investment, and the expectation of reward warrants their synthesis to explain regular as well as irregular, subjective, and countercultural forms of social mobility. It is necessary and possible to do so because "aspiration studies [are located] within that class of theories which presuppose individuals make efficient choices to maximize their satisfactions and minimize their dissatisfactions" (Sherwood 1989: 61).

In this light, the two concepts are linked to rational-choice theory and utilitarianism and to even more remote sources in seventeenth-century continental rationalism and the Cartesian cogito. They are also associated with the orientations of American Pragmatism, a philosophical movement that has been described as "a system of humanist metaphysics in which the idea of progress is built into human existence and with the notion that human beings can facilitate its operations" (Obono 2008: 238). Thus, Sherwood's pragmatist framework not only reinforces the idea that individuals invest time, effort, and money in line with their aspiration—defined as "any goal an individual is willing to invest in beforehand" (Sherwood 1989: 62)—but that they acquire capacities and competencies that increase their social capital in the course of doing so.

This is the framework used in this chapter to analyze the behavior of Nigerian migrants in the third phase of transatlantic migration. In the Nigerian setting, the construct of social capital facilitates thinking and our understanding of how people acquire and apply skills and knowledge as well as how *created social networks* are transformed into viable and durable relationships that provide or enhance value. The concept relates imaginaries of mobility with imaginaries of aspirations and "connections" in an individual's bid to maximize satisfaction and minimize stress in a global universe. It helped us derive the perception of benefits from individual- or corporate-level strategic alliances within the fluid milieus of Nigerian society, which approximates a "risk society." By framing it as an aspect of network discourse and political economy, we have explored here the local uses and appropriations of the term not as a residual theoretical category but as an active part of a compelling body of options calculated to aid the attainment of personal and collective goals and aspirations.

Conclusion

In Nigeria, connections confer privilege and impunity in a prebendalist regime that is characterized by patron-client relations rather than due process in the provision of basic services. It paves the way for one's entry into networks of trust, a compromise that may occasionally shield criminal activity and subvert due process. For this reason, establishing and maintaining connections have become agentive and imperative in Nigerian society. There has been a loosening of ties to the family and social heritage in the perpetuation of these connections. The space for participation is expanding for many persons who previously had limited access to resources, while the space is at the same time shrinking because the cosmopolitan elite is being threatened by competition that has become more intense.

As a consequence, there has been an increase in both innovation and deviation. Nigeria is witnessing the historic emergence of a class of entrepreneurs that will alter the basis, nature, norms, character, and consequences of the idea of connections. Whether in the sphere of religious leadership, corporate enterprise, or political rule, the idea that connections are indispensable to personal or professional growth is ingrained in the public mind and central to professional commentary in sub-Saharan Africa. For this reason, the effects of social mobility are better represented within frameworks that associate with connections and the acquisition of social capital.

The discussion in this chapter identified a third wave of migrant outflows cast against the historical settings of two previous patterns of forced emigration. This transition was explained as being accompanied by a reverse mercantilism in which the outward push was inspired by the quest for long-term connections. In the process, African transatlantic mobility, like the African transatlantic aspirations behind it, appears conditioned by the same mercenary worldview that once triggered slavery and colonization. These two patterns were products too of the ideational aspects of social capital. African emigration and mobility are currently legitimized by perceptions of interracial unity as a global value relative to the perceptions of difference that were behind the first two forms of historical migration.

The Ajala motif has enabled metaphorical analysis of processes underlying this third wave. It was used here as a semiotic device to organize early emigrant motivations although it does not explain the perversions and subversions that began with more criminal forms of emigration in the 1980s. This departure was an errant form of mobility that is not supported by norms of propriety in the communities of origin. This is what we referred to as the mobility of norms—a bundle of communal aspirations conveyed by migrants who project the identity of their home community wherever they go.

Ajala was celebrated precisely for these reasons, not merely because he undertook unlikely travels "all over the world." They were improbable because of the circumscriptions of time and the logistic burden involved, and were important because he personified an emergent regional, societal, and/or national aspiration. His travels were the congealed portrait of what society wanted at the time. He was an exemplar. This is why modern Nigerian "frequent flyers" are anecdotally referred to as "Ajala," demonstrating the fit between his exploits and a visceral need to replace the dominant sedentarism that had been the main attribute of the older social order.

It is noteworthy that among the Igbo, who occupy a landlocked territory in southeastern Nigeria and are being pushed outward by restrictive ecology, there is no parallel to Ajala. This may be because Igbo outmigration was a strategic response to limited space. With high reproductive motivation and a caste system in parts of the Igbo hinterland, migration was established early as a way of life. The absence of an Igbo equivalent to the Yoruba Ajala therefore conveys the same significance of mobility. What the Yoruba expressed through the celebration of Ajala, the Igbo achieved by routinizing mobility and establishing networks that fostered subsequent migration.

These two groups are Nigeria's second and third largest ethnic groups, respectively. The routinization of mobility as a means of establishing ties, particularly in Europe, is a major asset that forms one dimension of individual and group social capital. The size of their populations makes Yoruba and Igbo migrants omnipresent in Nigerian society and in many other countries. Consequently, their norms of mobility and social-capital formation have diffused and intensified patterns of mobility among concentrations of minority groups in southern Nigeria.

Notes

1. *Tokunbo* is a Yoruba name commonly given to children born overseas (especially in Europe or the United States) but, in modern usage, has been extended to include secondhand cars and vehicle and electronic parts that, since the early 1980s, have been imported or smuggled into the country from Europe (notably Belgium). These cars and equipment are widely perceived to be more reliable than even brand-new products whose life spans are thought to be much shorter.
2. See also Popoola *et al.* (2011: 58), Omoniyi (2010: 244), Ogunbodede (2008), and Ajayi & Dosunmu (2002).
3. Many Nigerians aver that Olabisi Ajala was a real historical personage but documentary evidence of this is scanty. This is unusual for someone of such global prominence who lived as recently as the middle of the twentieth century. In the absence of stronger historiographic evidence, musician Ebenezer Obey's song is the most credible extant account and can be trusted because it was conducted in the praise-singing tradition of the Yoruba bard. Songs of this nature are never sung in praise of purely fictional characters. For reasons of patronage and marketing, the genre tends to celebrate the life of real persons.
4. We are grateful to Akinyinka Akinyoade at the African Studies Centre, Leiden, for his insights on Ajala's mode of travel although we have reservations about its functionality. Our arguments in the main text explain this position.
5. Amebo, a character played by Veronica Ibidun Allison (b. 1941) in the Nigerian comedy, *Village Headmaster*, that was aired on Nigerian television from 1968 to 1993, was an *olofofo par excellence*. An *amebo* is defined as "a person, mostly a woman, of Nigerian descent, of questionable, cowardly character and low morals, esp. one who delights in EXTREMELY idle talk; an unproductive individual; a silly twat" (Ness 2009).

References

Adepoju, A. (2010) "Introduction: Rethinking the Dynamics of Migration within, from and to Africa," in: A. Adepoju (ed.), *International Migration Within, To and From Africa in a Globalised World.* Legon-Accra: Sub-Saharan Publishers, pp. 9–46.

Ajayi, A. B. & O. O. Dosunmu (2002) "Environmental Hazards of Importing Used Vehicles into Nigeria," in: "Proceedings of International Symposium on Environmental Pollution Control and Waste Management 7–10 January 2002," Tunis (EPCOWM'2002), pp. 521–532.

Bankston, C. L. III (2004) "Social Capital, Cultural Values, Immigration, and Academic Achievement: The Host Country Context and Contradictory Consequences," *Sociology of Education* 77(2): 176–179.

Bourdieu, P. (1985) "The Forms of Capital," in: J. G. Richardson (ed.), *Handbook of Theory and Research for the Sociology of Education*. New York: Greenwood Press, pp. 241–258.

Ecclestone, K. & J. Field (2003) "Promoting Social Capital in a 'Risk Society:' A New Approach to Emancipatory Learning or a New Moral Authoritarianism," *British Journal of Sociology of Education* 24(3): 267–282.

Fukuyama, F. (2002) "Social Capital and Development: The Coming Agenda," *SAIS Review* 22(1): 23–37.

Livingstone, D. & R. Harrison (1981) "Meaning through Metaphor: Analogy as Epistemology," *Annals of the Association of American Geographers* 71(1): 95–107.

Maia, J. M. E. (2011) "Space, Social Theory and Peripheral Imagination: Brazilian Intellectual History and de-Colonial Debates," *International Sociology* 26(3): 392–407.

Ness (2009) "Amebo—Yes, I'm Talking to You', *Bella Naija*" http://www.bellanaija.com/2009/04/22/amebo-yes-i%E2%80%99m-talking-to-you

Obono, K. (2008) "Reproductive Change and Communication among Mothers and Daughters in Ugep, Cross River State, Nigeria," Ph.D. Thesis, University of Ibadan.

Obono, O. (2008) "Ideology and the Global Context of Nigeria's Population Policies," *Nigerian Journal of Economic and Social Studies* 50(2): 235–256.

Obono, O. (2010) "The Coming African Exodus," *Business Day* (November 29). http://www.businessdayonline.com/NG/index.php/analysis/columnists/16410-the-coming-african-exodus

Obono, K. & O. Obono (2009) "Media Exposure and Reproductive Behaviour Change among Generations of Adolescents in Ugep, Nigeria," *Journal of the Nigerian Sociological and Anthropological Association* 7: 81–99.

Obono, K. & O. Obono (2010) "Gender and Reproductive Communication in Ugep," *Journal of Communication and Media Research* 2(2): 67–76.

Odeyemi, D. (2004) "Between President Obasanjo and 'Ajala Travel'," accessed December 1, 2004: http://nigeriaworld.com/feature/publication/odeyemi/120104.html

Odukoya, O. O., T. A. Arowolo & O. Bamgbose (2000) "Pb, Zn, and Cu Levels in Tree Barks as Indicators of Atmospheric Pollution," *Environment International* 26(1–2): 11–16.

Ogunbodede, E. F. (2008) "Urban Road Transportation in Nigeria from 1960 to 2006: Problems, Prospects and Challenges," *Ethiopian Journal of Environmental Studies and Management* 1(1): 7–18.

Oladeji, M. O. (2010) "Perception of Smuggling among the Yoruba People of Oke-Ogun of South Western Nigeria," *Internet Journal of Criminology.* http://www.internetjournalofcriminology.com/Olaniyi_Perception_of_Smuggling_South_Western_Nigeria.pdf

Omoniyi, T. (2010) "Language and Postcolonial Identities: An African Perspective," in: C. Llamas & D. Watt (eds.), *Language and Identities.* Edinburgh: Edinburgh University Press, pp. 237–247.

Popoola, O. E., A. A. Abiodun, O. T. Oyelola & L. N. Ofodile (2011) "Heavy Metals in Topsoil and Effluent from an Electronic Waste Dump in Lagos State," *Journal of Environmental Issues* 1(1): 57–63.

Portes, A. (1998) "Social Capital: Its Origins in Modern Sociology," *Annual Review of Sociology* 24: 1–24.

Portes, A. (2000) "The Two Meanings of Social Capital," *Sociological Forum* 15(1): 1–12.

Portes, A. & P. Landolt (1996) "The Downside of Social Capital," *The American Prospect* (May–June): 18–21.

Sherwood, R. A. Jr. (1989) "A Conceptual Framework for the Study of Aspirations," *Research in Rural Education* 6(2): 61–66.

Smith, D. J. (2008) *A Culture of Corruption: Everyday Deception and Popular Discontent in Nigeria.* Princeton: Princeton University Press.

van den Berg, M. (2011) "Subjective Social Mobility: Definitions and Expectation of 'Moving Up' of Poor Moroccan Women in the Netherlands," *International Sociology* 26(4): 503–523.

CHAPTER 12

Connecting "Ourselves": A Dogon Ethnic Association and the Impact of Connectivity

Walter E. A. van Beek

Introduction: The Office

Connectivity is a state of mind, also for Mamadou Togo, the president of *Ginna Dogon* (Big House of the Dogon), the ethnic association that represents the Dogon people of Mali. He happily receives me in his government office in Mali where he has a senior position in the Ministry of Foreign Affairs. It could be said that there are at least three types of offices for Malian civil servants. One has an office with very sparse furniture, just dusty filing cabinets and a few books that are quite clearly never consulted or even moved, and some folders on a desk that demand attention. People walk in and out of the door and the person whose office it is chats with others, talking on the phone more than writing and certainly talking more than reading. The simple sturdy metal chairs for visitors are stacked on top of each other, the paint on the walls is old, and there is no air-conditioning. The second kind of office is that of someone who has made it: a sumptuous air-conditioned office dominated by a large conference table with impressive but surprisingly uncomfortable chairs, a splendid desk that is almost completely empty but is well polished and clean. In the corner of the "salon" is a flat screen TV as all football matches have to be monitored officially. This office has white imitation-leather couches in the office itself but also in the waiting area attached to it. These are for the guests who will need to wait till the official is ready to greet them. The third type of office is like Togo's: a moderately roomy air-conditioned office with some comfortable chairs facing the central element, the desk that is covered in loose-leaf files that are all stacked on top of each other. But there are also files and books all over the office. Work is clearly being done here. This is the office of the public servant who takes all the decisions that are pushed up from below and delegated from the highest level downward, and who then respectfully disguises his own decision as an initiative of his superior. Each well-functioning department has a few of these persons, and they are usually at a level just below the political top. The lowest-level office, our first one, houses someone who, in the old Soviet regime's discourse "pretended to work like the state pretended to pay him." The second is the minister who has staff to do the work for him and whose job it is to keep up his connections with other patrons and the head of state. The third type falls somewhere in between; he is paid and he does the work.

Mamadou Togo has had a long and distinguished career in the public sector. He is a Dogon himself (Togo is his *tige* or village-based family name) and as a Dogon he has moved up in the Malian public service, mainly through the diplomatic ranks. His last posting was as Malian ambassador to the United Kingdom and he clearly relished his stay in London and speaks fluent English. Mamadou Togo is the force behind the *Association Malienne pour la Promotion et la Protection de la Culture Dogon* called *Ginna Dogon* or "Dogon Big House," which is the usual indication of a lineage house. He is proud of his organization and also of his Dogon identity: the two are joined at the hip. The association has not always been under his leadership but he more or less rescued it from stagnation and much of the story that I tell here is the direct result of his initiatives and organizational acumen. During our first interview, Togo was continuously receiving calls, making calls, responding to people who walked into his office, looking up papers, searching for statutes and the minutes of meetings, all the while telling the story of *Ginna Dogon*. He is continuously connecting, he is Mister Connectivity himself.

Ginna Dogon

Ginna Dogon was founded in 1992 in the early days of Mali's democratization process when the "troubles" were over and Moussa Traoré had been ousted. The interim government was preparing for elections when some Dogon notables—Togo's predecessors—suggested organizing an ethnic Dogon association. In the "Moussa days," the political credo was centralization, and local and regional organizations were frowned upon. Malians were then considered to be Malian first and Dogon, Bambara, or Soninke second, or not at all.

Not only Ginna Dogon but quite a few other associations sprang up in West Africa from the 90's onwards. For instance in Cameroon a similar development took place. In that country the regime change from Ahidjo to Biya facilitated grass roots movements which had been prohibited, even suppressed during the centralizing Ahidjo regime, both regional movements and local ones, even including local draughts clubs. Philip Burnham describes the genesis of a Gbaya association MOINAM in 1994, during more or less the same period.

(Burnham 1996: 139–143)

The *Mouvement d'Investissement et d'Assistance Mutuelle* with its clever name sounds like a *tontine*,[1] but in Gbaya *moi nam* means "gathering kin" (*Ibid.* 1996: 140). For insiders this had a definitely ethnic ring to it.

When the idea for *Ginna Dogon* was put before the interim president, Amadou Toure, in 1992, he asked whether it would be a political or cultural association. Malian politics had never encouraged ethnic politics and the idea was to have "real" parties with a national political program.[2] Also, the ethnic divisions in Mali were too skewed with its large majority of Bamana speakers to render ethnic parties viable. This was to be a cultural association and the first statutes were drawn up for an association aimed at "protecting and promoting Dogon culture." The first documents did not define "culture" in the broad anthropological sense, but interpreted it as meaning "dress and language." So when the first *Assemblée Générale* was held in Bamako in 1992 they defined the goal in Dogon terms, so as to preserve *tèm* and customs as well as promote them. *Tèm* is a Dogon term meaning "found," that is, all items of behavior and objects handed down by previous generations with the inevitable implication of "ancientness." What exactly *tèm* comprised was an issue for later debate. During the first assembly, the board was staffed with four *antennes* or regional offices in the major towns in the Dogon area.

One major influence throughout has been tourism (see Figure 12.1 for an impression). Dogon country is the top destination for international tourists in West Africa, and the heartland of the Dogon with the Bandiagara Cliff that was designated as a World Cultural Heritage Site by UNESCO in 1989. This recognition was significant and was one of the triggers for the *Ginna Dogon* initiative. Protecting the *tèm* for the Dogon at that time did not mean protecting it from the impact of tourism. On the contrary, it meant opening up the area to tourism, to facilitate access and especially to spread it more evenly over the *pays Dogon*. The old regime, with its Marxist leanings, was based on centralized parastatals, one of which was the *Société Malienne de l'Exploitation des Resources Touristiques* (SMERT), a name that reflected the top-down view of tourist resources (read "people") as government property. SMERT coordinated, facilitated, and organized all the tourist flows as well as cultural performances across Mali, but especially those in Dogon villages where tourism was concentrated. "We are in the hands of SMERT" people explained when I arrived in Sangha in 1979 for the first time. With regime

Figure 12.1   Tourists and their masks, Tireli 2010. (Photo: W.E.A. van Beek)

change and market liberalization in the 1990s, SMERT was disbanded, the state hotels were sold to individual entrepreneurs, and local, regional, and international tour operators appeared on the scene.

Tourism has been a major source of revenue for Mali and the Dogon, who generate the tourist appeal. And this is what the *Ginna Dogon* wanted. Most of the founders—and the present board as well—do not stem from Sangha and decried the central place that Sangha had in tourism as it is a cluster of ten villages but they are by no means representative of the Dogon area. Their central place was generated by a long history of both colonial openings in the area, and as a result of a specific and tendentious French ethnography.[3] Tourists came to Sangha as the epitome of the Dogon, and to the Dogon as the icon of the *savant primitif* (van Beek 2003). The first priority of the *Ginna Dogon* was thus to spread their revenue by facilitating access to the area. They were not particularly concerned about the influence of tourism on Dogon culture, and considered it to be under a much greater threat from the influence of "modernity," that is, the encroaching cultural dominance of the North. Globalization was more on their mind than "touristization."[4] Clothing and fashions remained the first signals of change but, being a male-dominated association at the time, they meant female clothing, mainly the blue cotton *pagne* that Dogon women still wore at the *falaise* in those days, and the same held for hairstyles. Not to belie their fears, wax-printed commercial clothing is now replacing the indigo cloth.

At the second *Assemblée Générale* in 1995, delegates came from all over Mali.[5] Ambajo Kasougué was the Minister of Culture then and the *Ginna Dogon* had full government support. He would later become the association's *Président d'Honneur*. Again, the assembly busied itself with filling its own ranks, and the *antennes* swelled in numbers. The majority of the delegates enjoyed having a titled post in the *Ginna Dogon*, so the assembly ended in a self-congratulatory mode. With changes in the Ministry of Culture, however, the association then seems to have gone into hibernation: there was no assembly in 1998 and the *Ginna Dogon* joined the ranks of sleeping organizations.

Togo arrived back from London in 2003 and took over the *Ginna Dogon*. He changed the board and called all those concerned to prepare for a "resurrection." On December 27 and 28, 2003, *Ginna Dogon* was reinstated in the *Palais des Congrès*, Mali's most prestigious conference center. As nobody can be dismissed, the organization is still growing. The new board installed a section for youth, one for women (times have indeed changed!), and, very importantly for Mamadou Togo, a *conseil des sages*, a council of 32 wise old Dogon men, aged 60 or older who originated from various villages, though some have lived in Bamako. One major change was that *Ginna Dogon* was no longer a city-based organization but to some extent rooted back in the village, both at the Bandiagara Escarpment and beyond. The elders mainly came mainly from the villages, and the delegates from Bamako and the regional offices toured the villages as well. Togo also managed to install outposts (*antennes*) in all the major cities around the world where Dogon lived, making up quite an impressive list of Dogon "colonies" in Paris, Riyat, Gabon, Abidjan, Burkina, the United States (mainly Washington), and the Netherlands.

Such an organization has to have a goal in terms of activities so Togo and his team set up two for *Ginna Dogon*. One was intellectual, while the other was related to performance. The *Apport Intellectuel*, as he called it, aimed at reappropriating Dogon ethnography. The main problem is not so much "foreign" ethnographic domination but the special character of Dogon ethnography. It has a special position in the anthropology of Africa, through the work of Marcel Griaule and his school. From 1938 to 1948 Griaule and his team studied the Dogon, which resulted in monographs on masks, language, social organization, and religion.[6] These descriptions fitted in well with the general picture of local African societies, but after 1948 a new picture of Dogon society emerged from the writings of Griaule himself. His 1948 book *Conversations with Ogotemelli* depicted a culture dominated by a complicated and convoluted myth of creation, which allegedly served as a blueprint for all facets of society from how to build a house to cultivating a field, to pottery making, weaving, basketry, smithing, drumming, and numerology. Posthumously in 1956, another completely different creation myth appeared on the scene, tying Dogon culture in with astronomy, with the creation of the whole cosmos, and especially with the double star of Sirius. Never before had anthropology seen anything like this and the *Conversations* book in particular was a runaway commercial success, eventually being translated into over 25 languages. Here was the truly exotic society, with all the aspects of social life firmly entrenched in myth and all the cultural details integrated into a coherent whole. Beautiful!

Too beautiful and too coherent thought many anthropologists such as Mary Douglas, Dirk Lettens, James Clifford, and others. In time, the Dogon case became a scholarly anomaly, standing out as the only culture in Africa with its complicated cosmology and complete regulation of life through mythology. For the public at large, the Dogon remained famous as an "exotic tribe," if not *the* exotic tribe. And their fame, whatever its grounding, gave an international boost to their art, sculpture, and masks, and the area's burgeoning tourism. Both surfed on this mythopoeia. New studies were late in coming but from the early 1980s onward, scholars from the Netherlands and Marseille initiated a new series of Dogon studies, partly to test the "Dogon anomaly," a movement I myself had taken part in. This resulted in a serious debunking of Griaule's post-1948 work, showing the Dogon to be an African ethnic group not unlike the others known by then.[7] There was no Dogon creation mythology to be found, at least not in the Griaulean sense, and no complete integration of symbolism or numerological systems, not even any traces. Others delved into Griaule's field notes and showed how much of Griaule himself had got into the descriptions of the Dogon, also on his central topic, namely, masks.[8] The new tales are naturally not as exotic and enthralling as the mythopoeia of Griaule and the new ethnographic descriptions still have to catch up publicity-wise with his mystification.

All these ethnographic antics mean that Dogon ethnography has, for a long time, been in the hands of the Europeans, first the French and other *Anjara* (white people) and later on the Dutch, the British, and the Americans. This had to change. "Everybody writes about the Dogon but the Dogon themselves," Togo said. So *Ginna Dogon* set out a program to redo the Dogon ethnography. Through Masters theses at the University of Mali and, if possible through Ph.D. projects, Dogon ethnography had to be reinvented. Aware of the writings of Marcel Griaule, the Dogon resented the mythopoetical and phantasmagorical twist his research had taken after 1948 and wanted to redress that.[9] Togo himself was aware of my 1991 refutation of his ethnography[10] but correctly surmised that refutation was not the same as a new, integrated description. He therefore wanted to use the council of elders as a screening committee for externally produced ethnographies. Griaule had already earned a "B-mol" for Dogon dissection in Togo's eyes, a low mark by anyone's standards, and other writers would follow, including me. I told him I would send my articles to him but they were mostly in English. This was no problem for him personally and he thought he could have them translated.[11] In fact, significant amounts of ethnography are now in the hands of the Dogon themselves, such as Tinta, Issaye Dougnon, as well as a series of informal publications on everyday Dogon life complied by tourist guides in cooperation with some writer-tourists.[12] But many of the new Dogon publications in English have still not yet found their way to the offices of *Ginna Dogon*.

Ginna Dogon also aims to have a library of all the films made on the Dogon, which is quite a project as over 150 films have been shot at the cliff face. I suggested that a bibliographical and filmographical project would be a good start and he agreed.[13] While talking to him about Dogon publications in print and film, it was clear that the Francophone productions had found their way into Mali more easily than the Anglophone ones had, though the latter are now much more numerous. One major problem that I foresaw was the generally poor condition of African libraries but the suggestion of tying projects such as this in with the National Museum in Bamako was accepted. After all, Mali can rightfully boast one of the best museums in sub-Saharan Africa.

Culture is an important issue in Mali, also politically. One reason for this is that the first democratically elected president of Mali, Alpha Oumar Konare, is an archaeologist who himself used to work at the national *Institut des Sciences Sociales*, after being Minister of Youth and Culture under the old regime. As one of his presidential projects, he decorated Bamako with a spate of new monuments, which greatly added to the visibility of culture in the capital. Second, Mali has quite a number of UNESCO heritage sites in comparison with most other African countries, with the Dogon area being just one of them. The others include Djenné's world-famous mosque and monuments in Gao and Timbuktu. Third, tourists come to Mali not for the usual African attraction of game, but for its culture, hoping for an authentic African cultural experience. Culture is therefore economically and politically important in Mali. And it is not only culture but the fact that through this recognition the cultural heritage of Mali has been connected with the global networks of culture and culture preservation. From 1989 to 1992 Konare was the first non-Western president of the International Commission for Museums (ICOM). This was a leading function that has put Mali and its heritage at the center of the world's cultural map. So through its links with the government, *Ginna Dogon* is finding its connections with at least one major cultural forum.

This internationalization of connections has resulted in an international-ization of people's view of self too. One of the changes is that, tied into this description of culture, *Ginna Dogon* also wants to produce culture. All over the Dogon heartland, masks dances are being organized, with tourism serving as a stimulus for development and change in Dogon masquerades.[14] Not only are dances organized for visiting tourists but many villages, especially at the southern end of the cliff, are also organizing their own dance groups that tour in Mali and beyond. Mask dance troupes from individual villages are invited by festival organizers in Europe and elsewhere and are taking their chance to appear on the international stage. And with acclaim too, one might say. I have personally accompanied three such groups in the Netherlands in the past decade. This was the market the association wanted to corner. No indi-vidual village could represent the whole Dogon region; this is now the task of *Ginna Dogon*. They wanted to organize their own dance troupe as the official Dogon one—or at least as an important one—for two reasons. This might of course have been a source of income but I think the main reason was that, with their definition of self and the character of their organization, taking the initiative in the most public arena of "Dogonhood" was a self-evident move. As the nodal point in the Dogon network, which is how they perceive themselves, it is their duty to represent their culture internationally.[15] And it is their very connection with the government, and through the govern-ment with the international cultural scene, that defines their organizational identity.

The national focus on culture and the production of Dogon culture are reflected in the main goal the association set itself: the organization of tri-ennial cultural festivals, And of all the ambitious projects of *Ginna Dogon*, this one bears the most fruit. I have never seen a *Ginna Dogon* dance troupe perform but the festivals have taken place, and on a grand scale. They are called *journées culturelles Dogon*, and avoid the term "festival," which they feel is loaded. Mali has more than its fair share of major festivals, like the *Festival du Desert*, but in Dogon eyes these happenings have become commercialized and dominated by professionals and artists from all over West Africa. As a "festival," the event would be taken out of their hands, which they did not want as they would then lose the cultural initiative vis-à-vis the Europeans. So they organized their *journées culturelles* in Koro in 2005, in Douentza in 2008, and the last one in Bandiagara in February 2011 (see Figure 12.2 for the Douentza installment).

Organization on such a scale is not easy, as they soon realized in Koro. The Dogon were, on the whole, enthusiastic, with most villages sending delega-tions to the festival. Koro 2005 was a huge happening and the people from my research village, Tireli, who participated described it as a wonderful gath-ering, but somewhat scary. Food and water were a problem due to the number

Figure 12.2 The elders of Tireli adhorting the masks in a tourist dance, 2011. (Photo W.E.A. van Beek)

of participants but, even worse, three people were wounded during the festivities in a melee generated by a cavalry stampede. Two of them later died. The people from Tireli did not go to the next one, which was too far away, but did attend the Bandiagara event.[16] The accidents did little to dampen the spirit of the *Ginna Dogon* enterprise, and they were not widely reported. Bandiagara 2011 and Douentza 2008 proceeded without incident, and *Ginna* officials vowed they had learned a lot.

The association's final ambition is to redress and resolve conflicts within Dogon society itself. They are very aware of the growing number of conflicts of the classic agriculturalist-cattle keeper type in the area between Dogon and Fulbe, but also increasingly between villages. Part of the problem is the increasing population density as well as a drier climate with more erratic rainfall, which has resulted in many of these conflicts being ecological in origin. *Ginna Dogon* has therefore created a commission on conflicts that tours the area to settle disputes. They once came to Tireli when I was there as this village has a long-standing problem with its neighbor, Ouli, on the right of access to fields.[17] In the traditional style of Dogon elders, they set up a meeting at the local school and then heard from the two parties in the dispute from Tireli and Ouli. Territorial disputes are almost impossible to settle, and in addition Dogon culture lacks any mechanism for settling disputes based as it is on equality and harmonious relations. So they did what all Dogon elders do. After hearing both parties, they put the blame on themselves for letting the situation arise: as elders they should have guarded the villages under their tutelage better. Then they asked both parties to respect them as elders, to forgive them, and to keep the peace among them. Typically, the evening ended in declarations of goodwill and intent, promises for betterment, and the elders left in a cloud of goodwill amid a flurry of good wishes. Nothing fundamentally changed between Tireli and Ouli and the situation is still tense, but at least there has been no violence since then.

The Three Connections

A modern ethnic association such as *Ginna Dogon* is generated, maintained, and promoted thanks to modern technologies. Three types of "bridges" are essential for such an ethnic association, namely, bureaucracy, communication technology, and tourism. The first is found in the trappings of bureaucracy, the social technology that shapes official bodies in Mali, as elsewhere in the world, and gives a general structure to *Ginna Dogon*, as it does with other similar associations. The structure itself, the positions, have statutes and a *Règlement Intérieur* and hold general assemblies with reports and voting. In short, the whole administrative blueprint has nothing to do with the way the Dogon organize themselves in their village community but everything to do with state bureaucracy. So the first connectivity is with the state, and links with the Malian state were numerous and strong; at the last general assembly in January 2010 the Ministers of Foreign Affairs and Public Communication (who is also a Dogon) were present, representing the President of Mali, Amadou T. Toure, who is the association's *Président d'Honneur*.

This type of globalization is fully hegemonic as no one in the history of *Ginna Dogon* ever thought of producing any other type of organization, for example, one closer to the Dogon way of doing things. But globalization generates localization as well, and so does the world bureaucratic blueprint. *Ginna Dogon* set out not only with an executive board, but with a huge one. At present the full *Bureau* has 35 members: the president has 4 secretaries, and so have the officers responsible for "*Arts et Culture*," "Organization," "Exterior Relations," "Communications," "Education and Scientific Research," and "Conflicts." Malians have appropriated this bureaucratic connection by creating a large number of posts, filling them with names, and publishing complete lists of employees wherever possible. An organization should create posts, not work. One reason why the association became less visible after 1998, I surmise, was that it had fulfilled its basic premise as a Malian official organization: it had given everyone a post, a title, and a place in the new organization, and when their business cards had been printed, its work was essentially finished.

For a European raised on Weberian notions of bureaucratic efficiency at least as ideal types, this is akin to putting the horse behind the cart, as such a large board would seem unworkable. The point, however, is that the board has already achieved its major goal. It is not work-efficient but socially efficient, as it constructs identity and links people up in a clear official structure. In short, it produces relations, it connects. Sociologist Pierre Bourdieu defines the various forms of capital as economic, financial, social, and symbolic.[18] Social capital is the network one can muster and the totality of the people one can address and use as a resource. For West Africans it is one's social capital that counts—the networks they can belong to and the people they come into contact. Boards are, without a doubt, a splendid hotbed of social capital. At the start of this chapter I described Togo's office in a slightly tongue-in-cheek manner to indicate his pivotal function. As a large board is not a working instrument but social capital, when something has to actually be done, the task falls to one person only. In the case of *Ginna Dogon* this holds as well, and it follows the Malian pattern too. Thus, the first connectivity—the apparatus—is appropriated by the culture in question, reshaping its content, goals, and impact.

The second connection is technological: the Internet, the (mobile) telephone, and print media. *Ginna Dogon* would never have existed without the mobile phone, and its revival coincided with the arrival of the cell phone. I have never had a conversation with Togo without him being interrupted by an incoming call. After all, no Malian—or African for that matter—ever turns off his/her phone. But the new media are crucial in other ways too. The Dogon area has never been well serviced by landlines and until the late 1990s the one landline in the cliff area was at the post office in Sangha, and this took incoming calls only. The mobile phone has rendered this outdated and changed the communicative landscape dramatically. Now, most Dogon villages are within phone reach, and the number of cliff-side phones is increasing rapidly, meaning that Dogon in villages have lost their communicative isolation. Figure 12.3, below, illustrates the Dogon solution to an intermittent reception of the telephone signal.

The connection of the local village, however, is first and foremost with the center of Mali and with Bamako, much more than with local villages. Most incoming calls are from abroad. The phone has thus multiplied intraethnic contact but in a nodal way, with the capital as the core of Dogon internal connectivity. This is precisely what *Ginna Dogon* does too as most of its bureau members are from the city, primarily Bamako. The new connectivity is redefining relations between the core and the periphery. The core area for the Dogon is still the Dogon area of the Bandiagara Cliff and its adjoining plateau and plains, but it is relating to itself now through the center of the

Figure 12.3 Phones waiting to be called, Tireli 2010. (Photo W.E.A. van Beek)

country. As the association promotes an ethnic discourse, the definition of core Dogon ethnicity is increasingly being undertaken in the city and not in the villages. The old core area is thus losing its discursive hegemony over the Dogon diaspora as the urban Dogon, who are also predominantly Muslim, take over the power of definition of what Dogon culture means and implies. It is emblematic to see Dogon notables in Bamako, dressed in full *boubou* as befits good Muslims, discussing the value of traditional Dogon dress and coiffure, items they left behind long ago, and extolling the virtue and meaning of masks they will surely never wear. In content, Dogon culture is a thing of the rural villages, but the discourse is theirs.

The Internet is still in its infancy in Africa though the association does have its own Web site, which is updated occasionally. African organizations operate by phone more than via the Internet or e-mail as computer usage is low, in fact almost nonexistent in the rural areas. The Internet is, however, crucial for the third bridge: connections with tourism. The association has oriented itself toward tourism since its inception and has aimed to open up the Dogon for tourists and to get a piece of the pie for itself. The present attitude is a little more guarded vis-à-vis tourism, trying to find a balance between crass commercialization and confident promotion, although the latter still dominates. Culture is something to show off, a notion that fits well in the performance orientation of Dogon society and religion but who directs the performance, and who invites the audience?

The liberalization of the tourism sector has been crucial for the Dogon, and one of the association's goals has been partly realized, namely, extending tourism to cover the whole area. After the centrality of Sangha, most of the cliff-side villages are now regularly visited by tourist groups that hike along the escarpment. This still leaves most of the plateau and the villages on the plains out of the tourist equation but these too are trying to generate their share of the (substantial) revenue from tourism. Flexible as tourism may be, this is hard to implement as the visual attraction of the cliff with villages huddled against it is undeniable. Here one crucial difference surfaces between the tourist attraction as such and the vision of the Dogon of their own culture. For the latter, like *Ginna Dogon*, the crux of the matter is their culture, with the mask performances as its very core. The setting is secondary, in fact a given. That may hold in the case of touring mask groups performing on a stage but it does not hold for tourists visiting the area: they want to see the masks in their proper setting and nowhere else. After all, they come for the *Pays Dogon* and for the splendid scenery with a spectacular culture, in that order. So this is one the association's goals that has already met its limits, as an increase in tourism to the villages out of sight of the escarpment is hard to envisage. The conservation of Dogon culture is a goal that many NGOs have made their own. For instance, the German Development Cooperation has set up ecotourism projects in Dogon country and the Malian state created the *Mission Culturelle* in Bandiagara in 1993, meeting the exigencies of UNESCO's nomination of the Dogon area as a World Heritage Site. One goal was, again, to spread the boons of tourism over the whole area and to engage the population in cultural and tourist activities.

Connection, Culture, and the Power of Definition

Connectivity, in the sense of being aware of one's connectedness, changes one's perspectives in three ways: of one's situation, of the outer world, but basically of oneself. This is what is happening in *Ginna Dogon*, through its *iournées culturelles*, its links with the government, and simply the fact of its organization. In connecting the Dogon among themselves first and later with the outside world, the organization has created a coherent vision of what Dogon society, culture, and its territory are, or at least should be. The fragmentation of Dogon culture, comprising at least six major dialects as well as large differences in expressive culture, is summarily acknowledged on the *Ginna Dogon* Web site but it has also relativized Dogon culture. Masks are defined at its core, notwithstanding the fact that most Dogon villages have no mask tradition at all, just as most fall beyond the itinerary of the major ambulant ritual of the Dogon, the *sigi* (van Beek 1992). So the first creation is that of cultural unity within the circumscription of the ethnic territory.

Through its nodal point in an organizational and discursorial network, *Ginna Dogon* is operating in a field of ethnic definitions. Connectivity leads to a new definition of ethnicity, as it offers a new perspective on one pillar of ethnic definition: culture.[19] Culture forms the main discourse but strictly within ethnic confines and, as such, this connectivity leads both to inclusion and to exclusion, inclusion as Dogon and the exclusion of all other Malians, the inclusion of those Dogon willing to tie themselves to the aims and claims of *Ginna Dogon*, and the exclusion of those Dogon that are trying to set up their own network with the outside world.

The discourse on culture is in seemingly primordial terms but through the processes of widening connections and the dispersal of the Dogon themselves, they have become a kind of "imagined community," one that has to be "promoted by making a few . . . diacritica highly salient and symbolic" (Barth 1994: 16). What exactly are these aspects of culture that serve as icons for the internal definition of self, results from exchange, performance and recognition, and also from the "power of definition"? *Ginna Dogon* vies for a part of this power to define what being Dogon is, and effectively uses the performances for that goal. This changes the notion of culture from the classical "integrated whole" toward a selection of spectacular and distinguishing items that appeal to the Dogon themselves and to the outer world—at least to that relevant part of the outside world that is deemed to be interested in Dogon culture. Dogon culture is now defined in a Heider-like triangular relationship, with *Ginna Dogon*, Dogon culture, and the external other, and between them are their own view of their culture, their appreciation of the relevant other, and the view they think the other has of their culture. Balance theory indicates that these relationships should be in harmony in order to be stable. If all signs are positive (as in Figure 12.4), the system is in balance, also if all are negative, but if one sign is negative, cognitive labor is called for to bring the assessments into harmony. In the case of *Ginna Dogon*, this means that the Dogon have a positive view of their own culture and their relevant others, so perceive the other to have a positive evaluation of Dogon culture too. This is the kind of self-image Mead called the "mirror self," and indeed this

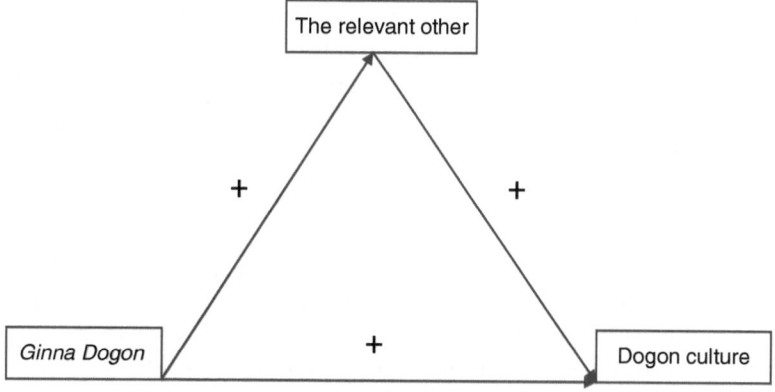

Figure 12.4 A balanced system.

connectivity-based view of culture involves an exchange of images between the "other" and oneself, that is, images of culture and thus mainly of cultural productions. But it is also an imagined exchange, as the flow in information through the connections is partial, skewed and depends on untested views, as there is neither discussion nor systematization. Urban-based *Ginna Dogon* members through their urban and professional careers, as well as Islamization, have changed more than they care to realize and any lingering doubt as to their cultural competence is quickly doused in the repeated insistence that they are "all Dogon."

So Dogon culture, which forms the basic discourse in this whole endeavor, is redefined by the arrows in the triangle: the perceived connections between the parties in the exchange. An ethnic association like *Ginna Dogon* simply has to use a unifying and effectively essentializing discourse, just as the whole notion of protecting and promoting *la culture Dogon* immediately and inexorably implies that one is talking about one culture, and then about a culture that exists independently of the observant, participant, and certainly of the reporter: a tribe independent of a scribe.

This process is selective and sometimes seems superficial. If prodded, culture is defined in its outward manifestations as language, dress, and coiffure, but in practice the main items are cultural productions more than daily externalities: the masks, the dances, and so on. What did the *journées culturelles* produce as cultural manifestations? The event held in Douentza in 2008 shows a wide variety of hybrid forms ranging from a traditionally dressed hunters' group to women in indigo cloths doing a circus act, from a Dogon sculpture to mask performances. A large part of the performances were either a parade of women carrying something, usually baskets, or troupes of *jeunesse* from a particular village performing their orchestrated dances. Most Dogon villages have one or two groups of youngsters who study dance moves and perform during yearly festivities.

From the variations in performances at these *journées*, two things emerged: first, the joy of performance itself, and, second, the performers' enormous cultural self-confidence. More than the particularities of dress—the indigo wrap has become almost standard for women—or of sculpted figures, the performance orientation is a crucial cultural feature in itself. The main Dogon rituals, such as the funeral and the mask dances, are exuberant, intense performances in which men perform in front of an appreciative female audience, while the women have their own means of showing off. Exhibiting cultural competence is an important value in Dogon society and this is apparent at the *journées* that are a cultural fair. The second is what I have dubbed "cultural self-confidence" in the framework of tourism (van Beek 2003): the idea that whatever one does as a Dogon, it will be interesting for others to watch. In short, this is the confidence that one's culture has an intrinsic interest. In my view, this characterizes Dogon culture more than folklore particulars. The sheer spectacle of wealthy Dogon, immaculately dressed in the gowns of West African power and the intricately designed women's dresses of the latest Bamako fashion, proudly performing at the central dancing ground, exudes an ethnic pride based on the rock of confident cultural appreciation.

So in *Ginna Dogon*, as one example of ethnic associations, culture is definitely back on the agenda, but in a connected fashion: culture is no longer self-evident but at the fore in people's minds and mainly seen through the eyes of the other, be they outsiders such as tourists or, far more importantly, urban dwellers who know enough to appreciate it but cannot or will not actually perform. And in true Dogon style, performance is a gentle competition between the villages, in this case between the various delegations from the villages. The *journées* are a cultural competition, the groups trying to "outculture" each other in a cultural arena that is reminiscent of their main rituals as well. For instance, mask dances are always judged by the visitors and compared with their own. So at these culture fairs, while defining Dogon "essentials" in folkloristic (and essentialist) terms, the main cultural repertoires reverberate in the performance. And through that expression of overt culture, hybridization is inevitable as they operate in what is a setting of very different connections.

Ethnicity, Connectivity, and Culture

This process of cultural ethnic definition runs counter to accepted anthropological wisdom. After the seminal *Ethnic Groups and Boundaries* (Barth 1969), the accepted basic anthropological model has been that ethnicity is about differentiation, is concerned with culture as the outcome of social interaction, is not fixed just as culture is not fixed, and is externalized in social interaction and internalized in personal self-identification (Jenkins 1997: 165). Writers such as Amselle, Bayart, and Barth have repeatedly stressed that content was subservient to difference, while others thought that cultural traditions were for a large part "invented" (Hobsbawn & Ranger 1983). There are reasons, however, to allow a larger place for "cultural stuff" (Barth's expression) and not limit ethnicity to the definition of a "taxonomic space" (Banks 1996: 132). Culture is more than a means of making a difference. Jenkins argues that "the cultural stuff is undeserving of neglect." If the main issue of ethnicity is the construction of differentiation, this inevitably entails a similar construction of similarity and, in order to be identified, there has to be something in common (Jenkins 1997: 168). This better focus on similarity, and thus content, also points a way out of the problem of infinite regression in ethnicity, the tendency to define the "own" group in ever smaller terms. Dogon can be divided into language groups, into village, and then the villages into halves, wards, and lineages. At what level does one identify oneself? That depends on the situation and one's goal but it is culture that makes one level of differentiation more germane than the others.

Did we overexert ourselves in denying the relevance of content-based descriptions of people's collective identity? Barth (1994:17–18) thinks we did: "The issue of cultural content versus boundary, as it was formulated [in *Ethnic Groups and Boundaries*], unintentionally served to mislead . . . Central and culturally deeply valued institutions and activities in an ethnic group may be deeply involved in its boundary maintenance." After all, the first law of social constructivism runs: "If someone believes a thing, it will affect what he or she does and it will therefore be real in its consequences" (Karner 2007: 16). But much more than border maintenance is at issue here, as the definition of Dogon is not made in terms of "otherness" versus neighboring groups, but of "proud selfness." In most of the ethnicity discussions in the North, the cultural card is played by minority groups vis-à-vis a dominant culture that runs a strong state. In both versions of that ethnic discourse analysis, the instrumentalist and the situational model, the ethnic definition is a tool to promote group interests depending on the circumstances (Fenton 2003: 84). The Dogon can never be accused of having a weak social identity due to the history of their interactions with the outside.

So we do not need to think in terms of "primordial" groups[20] when we recognize the importance of a group's self-definition in terms of culture. Burnham (1996: 155) writes: "We should not throw out the concept of ethnicity with the bathwater of modernization theory."[21] People may invent or reinvent their traditions but in the end they do have a "cultural logic" (Burnham), a "symbolic reservoir" (McEachern), and a "cultural reper-toire" (Bayart) at their disposal to operate with as part of their cultural competence.[22] In *Ginna Dogon* we perceive that when culture becomes a "connected" culture, as in any form of modernity, we all become, in Lévi Strauss's terms, cultural *bricoleurs*, reassembling bits and pieces that are famil-iar to suit our needs and, in so doing, fabricate something new that looks familiar enough to be considered deeply old. Tradition, as Pascal Boyer has analyzed, is a form of emic authority, part of a culturally acceptable com-munication leading to a comfortably recognizable truth (Boyer 1990). For whatever is performed, also for *Ginna Dogon*, it has to be recognized by the others as Dogon. So cultural content in an integral part of ethnicity, even for Barth (1994: 20): "We need to discuss not just the relations and reproduc-tion of social categories but the conditions for the perpetuation of cultural traditions and ways of life."

Those cultural competences often imply quite mundane forms of rea-soning and self-evident forms of representation, as "people have very rich repertoires of cultural knowledge which are continuously being implicated in the construction of social practice" (Burnham 1996: 162). Fenton (2003: 194) summarized the ethnicity discussion in three questions: "Are ethnic groups real, are ethnic groups corporate, are ethnic 'motives' calculated?" The first has been answered: they are constructed but the constructions are very much present. In short, they are "real constructions." The second question points to processes such as *Ginna Dogon*: ethnicity has become a product of connectivity, leading to conscious reflections and pervasive self-definition. Eriksen (2002: 129) argues, discussing processes of modernization and its "stages in ethnogenesis": "The emerging cultural self-consciousness or reflexivity brought about through these very processes, has also inspired the formation of ethnic identities stressing cultural uniqueness." And cultural uniqueness has to be shown. The third question, calculation, addresses *Ginna Dogon* as well. The association is highly self-conscious as well as planning ori-ented and uses culture in an instrumental way, as in the case of the dancing troupe. But that seems to be harder to realize than the assessment of just being Dogon.

We should not, therefore, dismiss the seemingly essentialist discourse of the Dogon about themselves as a kind of false reality: it is based on recog-nizable cultural processes and reconstructed into an even sturdier discourse. In this, informants should be taken seriously: they are now becoming col-leagues as well. There is no way in which I can speak about Dogon culture with the Dogon themselves and the *Ginna Dogon* leadership, without using a content-oriented discourse. Even if I may privately analyze Dogon culture as heterogeneous, a historical amalgam or a colonial product, I have still to realize that their discourse is dominant in their own view, as it is to some extent in the political arena, and fully so in the tourist scene. People think not of an empty form, they think of content, especially when considering identity, and thus ethnicity. And any discourse not only has to be generated, it also has to be recognized and accepted. Writing about a Gbaya ethnic association, Burnham (1996: 159) noted:

> Progressive restructuration of Gbaya ethnicity made use of the cultural materials to hand and was accomplished within the framework of the already existent Gbaya ethnic label, this more socially pervasive identity did not seem novel to the Gbaya actors concerned and has continued to be viewed as deeply rooted in tradition.

"Tradition" has long been identified as an *argumentum ad autoritatem*, the stamp of perceived antiquity on yesterday's cultural *bricolage*,[23] but it is still a powerful and convincing discourse that connects people by connecting them to a distant, imagined, and thus immensely powerful past. Also, some ethnic identities demand much more cognitive effort than others. The Gbaya that Philip Burnham talks about are much more heterogeneous, agglutinated, diversified linguistically and politically dispersed than the Dogon could ever conceive of being. Compared to them, Dogon culture is positively monolithic, even if most Dogon never dance a mask, speak any *sigi so*, or will ever participate in the *sigi*. The level of ethnic construction thus varies, which makes the discourse on tradition in some instances much easier to generate and to accept than others. And I have never encountered a Dogon for whom the discourse on Dogon culture was strange or uneasy, though all recognize that the culture has varied and changed. Everyone maintains that it is still Dogon culture and that they can recognize it as such.

A final thought—of the many connections that make up Dogon ethnicity, I am one as well, as are the many other anthropologists who have worked in the area. Our descriptions of Dogon cultural practices are based on the bedrock of any fieldwork, the field relations one has, the kinds of roles, conceptions, and performances that form our presence in the field. The notion of connectivity highlights the two-sided aspect of this link, as not only am I connected to me, giving me my insight in Malian society, but they are also connected to me, giving them another angle on their own culture through my eyes. The triangle of perceptual balance always works both ways, and forms the main mechanism by which one's culture is appreciated, and thus formed and formulated. In the end, connectivity is what anthropology is all about, as through our own connectivity we form an image of "them"—generating or constructing knowledge about the other—and at the same time also about ourselves, a cultural connection that is at the core of our profession. Trying to speak for our informants, or even when trying to have the "third voice" heard, we are inevitably part of the connectivity of our informants (and new colleagues). Our connecting and mediating function should not discourage us from making definite statements about the other. Our heightened awareness of our own connecting role in constructing knowledge and insight should generate some modesty but at the same time can lead to self-assurance. The bridge between cultures has a certain form, structure, and even content, which thoroughly informs and colors our writings, and has long been recognized to be also us, but it still is a bridge and we know it better now. So the cultural competence the Dogon feel should be matched by our confidence of description. We may be a constructed bridge, but a connection we are. *Ginna Dogon* is such a bridge too, shaping, coloring, and changing both sides of the river that it is connecting and filling that definition of "self" with great gusto and confidence.

Notes

1. This is the Francophone African term for a rotating savings and credit association.
2. The same still holds for Senegal. Ethnic associations are in principle not welcome in the Senegalese political context but, nevertheless, exist as homeland or urban-rural associations. For instance, *Ndefleng* is the association that unites all Serer, a minority group vis-à-vis the Wolof. Its main goal is rural development and it operates projects and *tontines*. In fact, most associations of this kind are either development oriented or political (Lucas 1994). Purely cultural associations, such as *Ginna Dogon*, are rare, which reflects the central position of culture in Mali and Dogon society.
3. van Beek (1991, 2004) and Lettens (1971).
4. Walter, Lemineur & van Beek (2008).
5. The *arrondissements* of Bankass, Koro, Bandiagara and Douentza, plus Dogon from Gao, Timbuctoo, Segou, Sélengué, Kayes, Kita, Koutiala, Bougouni, Yamfola, Niono, Markala, Cikasso, and Koumanton.
6. The team consisted of i.a. Debora Lifschitz, Michel Leiris, Denise Paulme, and, throughout her life, Germaine Dieterlen.
7. van Beek (1992, 2005), Hollyman & van Beek (2001) and work by Jacky Bouju, Eric Jolly, and Laurence Douny.
8. Anne Doquet, Polly Richards, and Barbara DeMott.
9. However, the lure of the exotic remains hard to resist. The printed materials of the *Ginna* accredited the masks with all kinds of cosmological interpretations and the Web site sports an article on Dogon "mysteries," meaning Sirius, long since thoroughly discredited. See Jansen & van Beek (2000), van Beek (2004), and Doquet (1999).

10. van Beek (1991, 2004) and Hollyman & van Beek (2001).
11. It has been sent and I am still awaiting comments. Also, on this article, which I personally gave him in his office.
12. Like Gindo, Wentholt & Molter (2005).
13. This is in the process of being organized.
14. Doquet (2011), van Beek (2012) and Richards (2011a, 2011b).
15. This project has not yet come to fruition.
16. The picture in this chapter that is not mine is from the second event, in Douentza, and was taken by pupils at the *École de Photographie* in Bamako. They are printed here with their permission (Photo 2).
17. See van Beek (2005).
18. Bourdieu (1977).
19. One other pillar is real or imagined territory. Here too for the Dogon, a case can be made for the connectivity basis of Dogon territory. See van Beek (2004).
20. Richards (2011b) calls this the *deme* model, using a genetic analogy.
21. In fact, this holds even more so for postmodern theory.
22. Christian Karner argues that Pierre Bourdieu's notions of habitus and doxa might well be used to distinguish between the surface characteristics of culture and its deeper structures. But that discussion exceeds the limits of this article, see Karner (2007: 56).
23. Boyer (1989) and van Beek (2004).

References

Banks, M. (1996) *Ethnicity: Anthropological Constructions.* London: Routledge.

Barth, F. (ed.) (1969) *Ethnic Groups and Boundaries: The Social Organization of Cultural Difference.* Boston: Little, Brown.

Barth, F. (1994) "Enduring and Emerging Issues in the Analysis of Ethnicity," in: H. Vermeulen & C. Govers (eds.), *The Anthropology of Ethnicity. Beyond "Ethnic Groups and Boundaries."* Amsterdam: Het Spinhuis, pp. 11–32.

Beek, W. E. A. van (1991) "Dogon Restudied: A Field Evaluation of the Work of Marcel Griaule," *Current Anthropology* 32(2): 139–167.

Beek, W. E. A. van (1992) "Becoming Human in Dogon Mali," in: G. Aijmer (ed.), *Coming into Existence.* Götenborg: IASA, pp. 47–69.

Beek, W. E. A. van (2003) "African Tourist Encounters: Effects of Tourism in Two West-African Societies," *Africa* 73(3): 251–289.

Beek, W. E. A. van (2004). "Haunting Griaule: Experiences from the Restudy of the Dogon," *History in Africa* 31: 43–68.

Beek, W. E. A. van (2005) "The Dogon Heartland: Rural Transformations on the Bandiagara Escarpment," in: M. de Bruijn, H. van Dijk, M. Kaag & K. van Til (eds.), *Sahelian Pathways. Climate and Society in Central and South Mali.* Leiden: African Studies Centre Research Report 79, pp. 40–70.

Beek, W. E. A. van (2012) "Masks in the Dogon Tourist Bubble: the Case of Tireli", in: W. E. A. van Beek & A. Schmidt (eds.) *African Hosts and their Guests. Dynamics of Cultural Tourism* James Currey, Oxford, pp. 37–57.

Bourdieu, P. (1977) *Outline of a Theory of Practice*. Cambridge: Cambridge University Press.

Boyer, P. (1989) *Tradition as Truth and Communication. A Cognitive Description of Traditional Discourse*. Cambridge: Cambridge University Press.

Burnham, P. (1996) *The Politics of Cultural Difference in Northern Cameroon*. International African Institute, London: Edinburgh University Press.

Doquet, A. (1999) *Les masques Dogon. Etnologie savante et etnologie autochtone*. Paris: Karthala.

Doquet, A. (2011) "Dogon Mask Festivals in the 21st Century," in: P. Richards (ed.), *The Dogon Now*. New York: Museum of African Art.

Eriksen, T. H. (2002) *Ethnicity and Nationalism. Anthropological Perspectives*. London: Pluto Press.

Fenton, S. (2003) *Ethnicity*. Cambridge: Polity Press.

Gindo, I., H. Wentholt & A. Molter (2005) *Het dagelijks leven van de Dogon*. Amsterdam: Mets & Schildt.

Hobsbawn, E. & T. Ranger (eds.) (1983) *The Invention of Tradition*. Cambridge: Cambridge University Press.

Hollyman, S. & W. E. A. van Beek (2001) *Dogon. Africa's People of the Cliffs*. New York: Abrams.

Jansen, J. & W. E. A. van Beek (2000) "La mission Griaule à Kangaba (Mali)," *Cahiers d'Etudes Africaines*, 40(4): 363–376.

Jenkins, R. (1997) *Rethinking Ethnicity. Arguments and Explorations*. London: Sage Publications.

Karner, C. (2007) *Ethnicity and Everyday Life*. London: Routledge.

Lettens, D. (1971) *Mythagogie et Mystification: Evaluation de l'oeuvre de Marcel Griaule*. Bujumbura: Presses Lavigerie.

Lucas, J. (1994) "The State, Civil Society and Regional Elites: A Study of Three Associations in Kano, Nigeria," *African Affairs* 93: 21–38.

Richards, P. (2011a) "A Systematic Approach to Cultural Explanations of War; Tracing Causal Processes in Two West African Insurgencies," *World Development* 39(2): 212–220.

Richards, P. (ed.) (2011b) *The Dogon Now*. New York: Museum of African Art.

Walter, O., P. Lemineur & W. E. A. van Beek (2008) "Tourisme et patrimoine au Mali. Destruction des valeurs anciennes ou valorisation concertée?," *Geographica Helvetica*, 4: 249–258.

CHAPTER 13

Intimate Strangers: Connecting Fiction and Ethnography

Francis B. Nyamnjoh

Introduction

Scholarship influenced by politics of exclusion has presented intra-African migrants in search of a productive and meaningful existence as an unbearable burden on those fortunate enough to be recognized and represented as locals, nationals, or citizens (Peberdy 2009; Neocosmos 2010). Locals feel resentment toward African "Others," whose presence is perceived as a threat, a danger, or an infection in need of urgent attention. Almost invariably, African migrants in African cities are perceived as epitomizing backwardness and as being on the limits of humanity, which must be contained if civilization and modernity are to carry the day. Citizens are instinctively expected to close ranks and fight off this "attack" by an influx of barbarians who do not quite belong and who must be "exorcized" so "insiders" do not lose out to this particular breed of "strangers," "outsiders," or "demons" who are perceived to have little but inconvenience and inhumanity to contribute (Landau 2011). This attitude is in contrast to fairer-skinned migrants from within and outside the continent who are believed to be higher up the hierarchy of "purity" of humanity often expressed in terms of belonging to racial, cultural, geographical, class, gender, and generational categories (Gupta & Ferguson 1992; Stolcke 1995; Geschiere 2009).

The perspectives and experiences of migrants themselves are absent. Studies reflecting them would require getting to know them as human beings, spending time with them in intimate circles, and developing research questions not of a slash-and-burn or rapid-appraisal nature but of an ethnographic type, with a focus on the complexities and contradictions of what it means to claim and deny belonging. The predicaments of migrants complexify once they arrive in their host country or community. Their reality is not as simple and straightforward as often suggested in the catalogue of stereotypes with which they are portrayed.

The aim of this chapter is to demonstrate how connecting fiction and ethnography can help bring out the perspectives of those neglected by mainstream scholarship. I have illustrated my argument of fiction and ethnography as intimate strangers with *Married but Available* (Nyamnjoh 2009), a novel based on a data set I used in scholarly journal articles such as Nyamnjoh (2005). Here, I reproduce excerpts from several chapters of *Intimate Strangers* (Nyamnjoh 2010), an ethnographic novel constructed from the same data set that contributed to the writing of the scholarly volume entitled *Insiders and Outsiders: Citizenship and Xenophobia in Contemporary Southern Africa* (Nyamnjoh 2006). This argument builds on my discussion of the negotiation of identity and belonging in fiction and ethnography (Nyamnjoh 2011). African fiction provides alternative and complementary ethnography of the everyday realities and experiences of Africans and their societies in a world of interconnecting local and global hierarchies, which is often not adequately captured in its complexities and nuances by the "ethnographic present" (Wolfe 1999) and its propensity for frozen and stereotypical perceptions such as those Chimamanda Adichie (2009) criticizes in "the danger of a single story." In addition, this chapter draws attention to the intricacies of being an intra-African migrant, in what are, perplexingly, not often considered diasporas in their own right (Bakewell 2008; Zeleza 2011).

By opting to contribute excerpts from several chapters of *Intimate Strangers* to the current volume of collected essays on global and local connections and interconnections, I am hoping to demonstrate that fiction has its place in social-science scholarship. Additionally, I am using these chapters to argue that mobility, connectivity, and connections by individuals are best understood as emotional, relational, and social phenomena captured in the complexities, contradictions, and messiness of the everyday realities of those we study. In conventional scholarly writing, even when such dimensions are recognized, the standard expectations of what constitutes a scholarly text do little justice to the multilayered, multivocal, and multifocal dimensions of everyday negotiation and navigation of myriad identity margins. I suggest that fiction as a genre is adapted to explore such realities and is complementary to scholarly ethnographic writing.

My study of connections is a study of insiders and outsiders, not as essences, birthmarks, or permanences frozen in time and space but as "intimate strangers" or as "frontier realities" (Kopytoff 1987). Being an insider or an outsider is always work in progress, is permanently subject to renegotiation, and best understood as relational and situational. Hence the need to understand the interconnecting global and local hierarchies—be these informed by race, place, class, culture, gender, age, or otherwise—that shape connections and disconnections, and produce and reproduce insiders and outsiders as political and ideological categories that defy empirical reality. To substantiate this point, my study documents the interconnection, inter-dependence, tensions, and conviviality among people with competing claims to places and spaces in contexts of accelerated and flexible mobility. I argue that fiction and ethnography are also intimate strangers: they complement each other in their flexibility, interdependence, and conviviality. The social scientist should be married to science but available to read and be informed by—and to write—ethnographic fiction.

What do we gain by connecting ethnography and fiction? Deeper social science for one thing. The lived lives of those who are not of the dom-inant race, place, culture, class, place or age are frequently swept to the sidelines of scholarship—and given voice in alternative spaces, such as music and literature (see Nyamnjoh 2011). Relying on ethnographic fic-tion as a legitimate source to inform investigations of social phenomena thus allows the researcher to embrace a wider variety of perspectives and provide more nuanced and accurate accounts and explanations, instead of uncritically reproducing dominant social constructs. Historian James Giblin (1999), who has his history students read novels by African authors, explains:

> Historians realized that many of the European writings which they use to reconstruct the African past—such as accounts by nineteenth-century mission-aries and travelers, for example—are . . . tainted by . . . notions of African infe-riority. . . . This realization . . . led historians to seek out alternative sources of information less influenced by European preoccupation with racial difference. These alternative sources include writings by Africans, . . . oral tradition . . . , the vocabularies and structures of African languages themselves, . . . physical artifacts [sic] uncovered by archaeologists. African art . . . [l]ike the other alternative sources . . . helps us . . . understand African history not from the standpoint of Europeans, but from the perspective of Africans themselves.

For sociologists and anthropologists, dipping into fiction can bring voice to silenced spaces and help science bridge rather than reinforce socially con-structed difference. Poverty-stricken, flat, and linear scientific explanations can become more multidimensional, more reflective of the complicities, contradictions, and compromises of everyday life.

Relating the results of ethnographic research in the form of fiction can make the work—in which society at large has certainly invested in various forms—available to readers beyond scholarly circles. I have found that even scholars reading research results delivered in a novel as opposed to a scholarly paper relate to the results in a more visceral way. Realities that might be difficult to explain in scholarly logic are felt and understood through the lived experiences of the characters in the novel. Intertwining these two different ways of writing is a way of bringing together worlds and worldviews kept apart by "scholarship" and its gatekeepers.

The ethnographic novel, some parts of which I share below to demonstrate the points mentioned above, is set in Botswana in Southern Africa. In the 1990s and early 2000s when I did my ethnographic research, Botswana was widely regarded as an island of prosperity in a continent of economic and political upheavals and uncertainty. Like neighboring South Africa following the end of apartheid, Botswana provided place and space for mobile Africans, marginal or otherwise, who were seeking fulfillment (Nyamnjoh 2006, 2010; Neocosmos 2010; Landau 2011). Such migrants compete for employers, resources, and other opportunities with those on the local mobile margins who feel more entitled to them as nationals and citizens. With a burning desire to survive and succeed, border-crossing migrants from Zimbabwe, Nigeria, Ghana, Kenya, Malawi, Cameroon, and a host of other African countries seem, however, more desperate and willing to be used and abused for much less than their local counterparts. This situation plays into the hands of employers, who have been quick to recognize the advantages of playing marginal insiders off against migrant marginal outsiders, and both of them off against the possibilities and limitations of the law and the state. These relationships are simultaneously distant and intimate, rewarding and alienating, material, and immaterial, and enhanced and contested by technologies. Connecting methodologies of fiction and ethnography enhances the possibilities for investigating, comprehending, and reflecting these interconnections with greater complexity and nuance.

The thrills and tensions, possibilities and dangers, and rewards and frustrations of social, cultural, and physical boundary making and boundary crossing are narrated in *Intimate Strangers* through the experiences of Immaculate, an outsider, a stranger or *makwerekwere*[1] from a fictional African country, fictional because I stress the universal in the particular, and the particular in the universal. Immaculate follows her fiancé to Botswana only to find him off in the United States and refusing to marry her. Immaculate is, however, determined to outwit victimhood. Operating from the margins of society, through her own ingenuity and an encounter with transnational researcher Dr. Winter-Bottom Nanny, she is able to earn some money as a research assistant. She learns how maids struggle to make ends meet and how their employers wrestle to keep them as "intimate strangers." Resolving to turn disappointments into blessings, she perseveres until she can no longer stand the repeated efforts of others to define and confine her. Through the relationships she forges with insiders, locals, and citizens, and with other outsiders, the reader is introduced to what it means to be an intimate stranger in a foreign country, competing with nationals and citizens and compounding their predicament. Hers is the story of the everyday, with the tensions of being and belonging that bring together different worlds and explore the various dimensions of servitude, mobility, and marginality.

The story provides an ethnographic entry point to understanding mobility, identity, and belonging from the perspective of migrants struggling to exist comfortably on the margins of their host communities. It invites the reader to experience some of the challenges Immaculate faced in Botswana, from acknowledging and questioning herself as *makwerekwere*, finding work, romantic relationships, and a troublesome existence to missing home and wanting to reconnect physically without necessarily disconnecting from her new, however precarious or tenuous, home.

Immaculate makes friends, draws on social networks past and present, and finds good but tedious jobs that allow her to make ends meet on the margins. As an outsider though, she is forced to acknowledge her own negative identity as a devalued foreigner and that of others like her, realizing that being different is unfortunately a reason for them to be treated differently. Finding herself in a stressful and at times horrifying love affair makes her long for her native Mimboland despite her extended stay in Botswana. The ambivalence with which she carries "home" is evident as she considers herself a stranger to her supposed homeland of Mimboland, having spent time and invested significantly socially and even economically in Botswana to be considered an insider, however tenuously. But she does not feel like a complete human being in Botswana. Dealing with one challenge after another, as a lowly regarded foreigner and as a woman, builds a person's character but also takes its toll. Despite her ambivalence, Immaculate eventually decides to go "home" to Mimboland. Just how naive is it to expect to be able to reconnect effortlessly? Does one underestimate how much one has changed during a prolonged stay away from home? These are open questions that reflect many a simplistic indicator of being (human) and belonging.

We can also think of anthropologists venturing out, as Immaculate did, to a new country, say, of fiction. The anthropologist becomes more of a stranger to anthropology and more intimate with fiction. After thrilling and also troublesome encounters, the anthropologist returns "home" changed, feeling enriched and connected to new realities. But he/she feels like a "stranger" to his/her home discipline. Should he/she expect to reconnect? These open questions consider many of the simplistic indicators of being (a scientist) and belonging.

Excerpts

Chapter 3: Being Makwerekwere

I don't know how it started, but a few days after losing my job, I was talking with Angel, when the word Makwerekwere became the centre of our talk.

"The term Makwerekwere leaves a bitter taste in my mouth," I told her, adding how often I had been reduced to tears by Batswana who called me this. "Could you tell me why Batswana tend to use this word in ways that stab and hurt?"

"One is Mokwerekwere, two or more Makwerekwere," Angel started in her soothing voice. She spoke like a Reverend Sister leading the Prayers of the Faithful.

She continued, "It is a shame we use the word the way we do, to refer to a particular type of foreigner from distant parts of Africa. Our neighbours from South Africa, Lesotho, Swaziland, they are not Batswana, but they are not called Makwerekwere either."

"Why not," I asked, bubbling with curiosity.

"Batswana seem to feel more comfortable with them than with other people from farther north—Zambia, Malawi, central, east or West Africa," Angel replied, searching the floor with her eyes.

"And so, if ever there's a Nigerian and someone from Lesotho, I'll find a Mosotho to be more like a cousin, more like family than I would the Nigerian," she went on.

"I think the whole thing goes back to this issue where we think you are here to take our jobs. We just tell ourselves, 'Oh! They are here to take our jobs.' We think these people are here because where they come from, things are bad. They came here because of our money, the Pula, and now they live more comfortable lives than we do, and so that's why conflicts erupt." Angel had an apologetic look.

"What about Zimbabweans, your immediate neighbours to the north? Are they Makwerekwere too?" I pretended not to notice the guilt in her eyes.

"With Zimbabweans, it touches my heart because I thought Zimbabweans are more of our sisters and brothers," confessed Angel.

"But Batswana, the way they treat them is like they are outcasts," she sighed.

"It touches my heart because Zimbabweans, we know why they come here illegally. They come here because of the situation in Zimbabwe. You can't stay in a place where there is no food while you know on the other side you could find food."

I nodded.

"The way I look at our border with Zimbabwe and other neighbours, there is something we can do about that because borders are man-made."

I again nodded, repeatedly, like a lizard.

"Take a look at the water sources for example. They used to unite us. But now, we say that side of the river is Zimbabwe, this side is Botswana, and so we shouldn't even share food and the water we drink," Angel shook her head in shame.

"Yet most of them come here not because they want to stay. They come here to do piece jobs and go back home. So why can't we allow them to do odd jobs that we often think are beneath us as locals?" I could see her face glowing with compassion.

"You tell me," I said, "Why can't you Batswana?" I wanted her to go on, as I found her words soothing, peaceful and promising of the world without borders I have always dreamt about. "Borders are our greatest killer," my uncle used to say, going down memory lane and detailing example after example of border conflicts that had eaten up sons and daughters of the soil.

Angel threw up her hands in resignation, before adding, "Normally in June, I think it's twice a year—June and December—the Immigration Police do what they call 'Clean-Up-Campaign.'"

She could see I was surprised by the expression. Telling me with her eyes that this was not what I thought, she proceeded to explain.

"This does not mean they collect litter like plastic bags, papers and tins. Oh no. They move from house-to-house, from workplace to workplace, to check all these Zimbabweans whom they see as litter. They deport them back to their country the way a person disposes of litter blown over the fence by wind." Angel covered her face with her hands, as if she was even then hearing the sound of human litter drop at the Zimbabwean side of the border.

I encouraged her to continue. It wasn't often to come across a local who was sympathetic and supportive, and when she told me she too was Catholic, I felt proud of my religion.

"Last year, the person who was supposed to supervise the teams doing the cleanup campaign was not in and my boss, the deputy director, called me in and said, 'Madam, the person who is supposed to do this and this is not in.' I said, 'What is this and this?' He replied, 'You know there is a cleanup campaign, we send all Zimbabweans back to Zimbabwe.'

"I said, 'What do you clean up?' I knew what he was talking about but I just said, 'What do you do? Am I to understand that all the litter we see in the streets is because of Zimbabweans, so we should dispose of it?'

"He explained, 'No, we move from house to house collecting all Zimbabweans everywhere—maids, garden boys and the like—, and sending them back where they belong.'

"I said, 'I don't think I'll be able to do that.'

"He said, 'Why?'

"I said, 'It's not because I want them to be here illegally. It's the way you people are handling this issue. And the people I'm supposed to go and supervise and work with, I'm going to have problems with them before I have conflicts with the law itself.'

"He said, 'Why?'

"I said, 'You can't talk of these people as if they are trash. We know they are here illegally and we know why. Maybe the best thing is just to say let's go out there and not clean up. Let's just check people who are here illegally and try to send them back home, and those who we know it's possible for them to have jobs here, we advise their employers to help them obtain papers to stay legally.'

"And the guy said, 'Are you Zimbabwean?'

"I said, 'Why do you ask? If I look at myself and you, you are more of a Zimbabwean than me.'

"He said, 'Why?'

"I said, 'Because you are a Kalanga.'

"He said, 'What are you talking about?'

"I said, 'I'm not going to lead this operation.'

"He said, 'You are going to do this job today. When you leave this office, you are going out into the field to supervise those people.'

"I didn't like the menacing tone of his voice. I told him, 'I'm not going. I'm going back to my office to do my day-to-day job. I can't imagine pushing people around and piling them into one congested lorry as if they are water melons.'

"He threatened sanctions, but I didn't budge. He could sanction me to hell. I didn't care. I just didn't care."

After her moving story, I stayed with the good Angel for a year. She taught me Setswana and initiated me to the ways and values of her land. She made me feel proud to be human, and living with her was like a year-long schooling in the dream that being different was no cause to be treated differently. I saw in her the embodiment of the gospel I had drunk all my life, of how we, regardless of race, place, creed or sex, are all children of the world, called upon to reflect the goodness of the Lord in the ways we live our lives.

If we had more people like Angel, I think the world would be a better place for us all.

Chapter 17: Madam and Maids as Intimate Strangers

Miss Amy Candlestick wore jean trousers and a blue T-shirt on which figured prominently "If Everything Was Everything." Following her after she opened the office door for us, I saw on her back, "HIV/AIDS: Save Africa."

. . .

"Have you ever employed a Motswana?"

"As maid, yeah."

"A Motswana or a Zimbabwean?"

"I think she is Motswana. She is the cleaning lady. She comes once a week. We don't have a live in maid, and we don't have a gardener."

"What's her name?"

"Priscilla."

"And what's her surname?"

"I don't know."

"It's not unusual. Most people don't know the second name of their maid."

"She doesn't know my name either. She came to us through another Canadian woman."

"How do you find her?"

"She is wonderful. She is a wonderful little girl. She is a young woman."

"And what is that?"

"I guess I am just comparing her. We've had two and the first one took advantage and made several hundred telephone calls when we went home, to the tune of P400. So I am just comparing her. She is friendly, she is trustworthy and she is hardworking. She is just very pleasant."

"The first one, if you recall, was she Motswana or Zimbabwean?"

"I think she may have been Zimbabwean, but I'm not sure. She was miserable and not happy."

"But she was hardworking."

"Yeah, until she figured out she could just make phone calls all morning," Miss Amy Candlestick said laughing. "And that's how we figured out that we couldn't just trust her. I had never figured somebody would do something like that."

"So what did you do when you found out?"

"Well, I confronted her and I said, 'I can't trust you anymore, and I am afraid I will have to ask you to please pack and that will be the end of our relationship.' I'm such a soft touch and she begged me to stay, promising she would deduct P50 a month until she had paid me back, and bla, bla, bla. We just paid the phone bill and were very careful not to leave money or our valuables lying around. You don't also want to tempt people by leaving money around. And then she found another job."

"It's interesting this issue of phones. It comes up repeatedly when I interview people."

"It never dawned on me because in Canada you don't pay for local phone calls. So it never occurred to me that it's a big deal, that you have to prevent people from using the phone. And I never figured that once they use it, they will want to use it ten times."

"Did you notice other things missing?"

"Well, we don't have a lot, and I never paid particular attention. But we have friends who employed her as well. And they noticed things like clothes going missing and her using things like cosmetics when she was there. Instead of working, she just took advantage of her work. In my circles if you employ people like that, we try to follow the Labour laws—try and make sure I pay Priscilla over Christmas and give her a Christmas bonus, make sure I give her sick leave—and she only works once a week—which is pretty generous. I find when people don't work in that situation, when they aren't familiar with that situation where they have those rights, they will take advantage of the employer. They will think you are a soft touch."

"You just mentioned a while ago that Priscilla is young and you live with your boyfriend or your husband. Doesn't it bother you that eventually the maid could become very comfortable and begin to look for ways of ousting you and taking over the ultimate object of your desire and love—your boyfriend or husband? Maids come in, they appropriate the kitchen, they appropriate the house and they clean everything and sometimes they even cook, and so they take everything from you except your husband, and even the bed they make it up and they lie on it and have their imaginations. They imagine themselves ultimate owners of what you hired them to take care of. Does it not occur to you as a woman who hires and fires the maid to watch out against losing the ultimate object of your attention and love? Does it worry you? Does it worry any people you know? If not, why not?"

Miss Candlestick was categorical. "No! It never actually crossed my mind and for a number of reasons. I feel my relationship is very solid, that it would never even cross my partner's mind. But he also has never really had much contact with her. She comes once a week for half the day. She often comes when we are at work. For the longest time, Wobble'd never met her. He leaves for work earlier than I do. And she is not an aggressive woman. Not that I have seen, but I think she just comes in, does her cleaning and hasn't been eyeing around and setting her imagination on fire. There's little we do that we would feel embarrassed about if a maid found out about it. We do our own laundry, we cook, all those things. If we hire someone, it is not like we fight over the house. We hire someone to do the mopping, the sweeping and that is it. I don't like the colonial attitude of coming in with a servant who appeals to me, but I do like the idea of having someone to mop."

"You say that she is not aggressive, but you know that the stereotypical presentation of the secretary is someone who is not aggressive, until the day the employer has a domestic problem and she offers him a shoulder to cry on."

"It never crossed my mind."

"I interviewed a maid recently who lived with a British family in a similar situation, where the husband did everything to have her."

"I think that has a lot to do with him rather than her and I think in my situation Wobble would never do anything like that because I believe he is not that kind of man. I've always gone for guys who are more interested in football and family than in playing hanky panky behind my back. It's a different kind of man who wants to exploit that kind of situation, who would take advantage of someone who works for him. The kind of man who sits with the secretary, this is the same kind of man who sleeps with his maid. I would be very surprised if I had that kind of situation from Wobble."

"What about you and say Batswana men? If you were to really find one who is decent in his approach, to what degree would you be tempted to pursue your fantasies?"

"I think the commitment that I have made in my relationship is such that the likelihood is virtually nil. It doesn't matter whether it is a Motswana, or a Ugandan or an American. It is not something I would do at this point."

"So not having a relationship with a Motswana would have nothing to do with the fact that Batswana are unfriendly?"

"No, I just wouldn't."

"What about with a Namibian?"

Amy Candlestick hesitated. "I wouldn't."

"And do you think there are other people here, in the Canadian community, who are interested in having relationships with Batswana?"

"I think the ones that I know are disinterested. They are yet to find the man who will make them interested."

"Of all of these dimensions, I think the greatest test of attitude towards people is the extent to which we go in creating and sustaining relationships with them."

"Just romantic relationships or...?" Amy Candlestick wanted clarification.

"Relationships in general, but the greatest test are romantic relationships for any community. The test of the relationship pudding is in the eating," Dr. Nanny explained.

"But I think for anything that came to me in that direction, it would have to be on a personality basis and not solely on their approach, the feelings you got and the culture they come from. No, I don't think I would have a problem. But knowing the HIV/AIDS rates here can be a deterrent.

"And if we go back to your previous questions, about husbands becoming not just comfortable but even close with maids, I wouldn't say those kinds of relationships represent necessarily open attitudes about how far one would go in venturing into another culture, because the husbands may just be seeing the maid as a sexual object and stopping at the bed . . . hmmm just like the maid might go after a husband just because of the size of his pocketbook." . . .

"Are you looking forward to leaving Botswana? Or are you looking forward to coming back?"

"No, I am not looking forward to leaving. But I am looking forward to being home in Canada. I am looking forward to travelling. It is nice to have a change. I just switched to this job in September and I really wish I had more time here with Aidswatch Network. I did not enjoy my time with WAP [Women Against Patriarchy]. It was a struggle from day one and I stayed longer than I should have. That tainted a bit my perceptions of work. I am not really anxious to leave, but I am looking forward to get home." . . .

Dr. Nanny desperately wanted to interview Miss Amy Candlestick's boss, especially as, like Johanna Salmon, Mrs. Birgit Rattlesburg was married to a Motswana. But she also didn't want the interview to take place the same day, partly because she was tired, partly because she wanted to take time off to digest the material she had gathered for the day, and more importantly, because she didn't want Miss Amy Candlestick to sit in and follow the interview with her boss next door. So she took an appointment with Mrs. Birgit Rattlesburg for The Queen's Arm, a popular bar with a touch of working class Englishness, for Saturday at 6 p.m.

Chapter 36: *Making Ends Meet as a Research Assistant*

Dr. Winter-Bottom Nanny was away for a long time. I had fulfilled my contract with her, by transcribing and sending via email the interviews, just as I had promised before she left. Satisfied with the work, she had paid me handsomely via the Western Union electronic money transfer service. It was the biggest amount of money I had ever handled. I proceeded to see how best I could invest it, only to run into difficulties, with my Zimbabwean boyfriend, Noway, and with a Motswana guy. I'm too scared to mention his name.

Exactly four years and six months after she left, I got a surprise phone call from Dr. Nanny, saying she was in Gaborone on a restitution visit. She had an autographed copy of her book for me, and could I meet her at the Gaborone Cactus Hotel at 6 p.m.? "Of course," I screamed with excitement. I wanted to see her new baby, the one I had helped to midwife.

Indeed, I was overjoyed to receive a signed copy of *Burdens of Womanhood: Being an Underling at the Margins*, which I couldn't wait to read, curious as I was, to see what she had made of my and the other accounts I had dutifully helped her gather and painfully transcribed verbatim. Although Dr. Nanny had told me that repetitive questioning was "the soul of ethnography," I was dying to know what she had been able to make of material collected through the boring practice of having everybody reply to the same set of questions.

Dr. Nanny could see excitement inscribed on my face when she handed me 2,000 USD as my share of what she termed "the generous royalties" she had been paid in advance for her book by her publishers. I didn't understand much about royalties, but I was pleased with the doors of possibilities that the money instantly opened up for me.

"With this money, I'm heading straight for Mimboland," I told her, amid hugs of appreciation.

"With Noway, I hope," said Dr Nanny, hungry for news.

"Noway is history."

She took a seat. "Tell," she said, like a master gossip.

"Story long, and time short," I tried to wriggle out.

But Dr. Nanny was her old stubborn self. "I'm in a hurry to go nowhere," she said, with a concrete look of you-seem-to-have-forgotten-the-patient-researcher-that-I-am on her face.

I gave in. "Then be ready to stay up all night," I told her.

She asked me to come with her to the poolside, where she ordered drinks, switched off her cell phone, and asked me to do the same.

"Now tell," she said, switching on her tape recorder.

"No taping this one," I warned.

She switched off the recorder.

And I began . . .

As I remember telling you several years ago, I met Noway on my way to Zambia. And the reason I was going to Zambia was to look for second-hand clothes to sell in Gaborone. But I didn't know that in Botswana, foreigners are not allowed to do that line of business. On going to Zambia, I forgot my residence permit, and at the border I had problems with the immigration authorities. I tried to phone some of my friends to copy my permit and fax it to me, but I couldn't reach them in time to continue my journey. One of the Immigration Officers took me to a lodge where I could stay the night. I hadn't a budget for that, so I had to use the money I brought for buying things from Zambia.

I was looking for somebody to phone back to Gaborone, when I met Noway. I asked him where I could phone, and he offered to take me there. I phoned Paul who faxed me the papers.

I was supposed to report to the Immigration Office in the morning at 8 o'clock. But Noway said, "No, don't go back because those people there are going to give you hell, better just avoid them." And I followed his advice.

The next day Noway came with a young boy, and they invited me for braai, but I said I was too tired to eat. They came again the following day. He asked me to sleep to be in a state to return to Gaborone, having advised me against continuing to Zambia, and against going back to the Immigration Office.

Anyway, that's how I came to know Noway. We travelled back to Gaborone together, in the company car they were using, and they dropped me off and we exchanged phone numbers.

From time to time he was coming to check on me, and eventually this led to a relationship.

In the beginning, he was a nice person, but he was staying with another lady I didn't know about, but we will come to that. He would come and check on me. I was staying with Christians, and they would not allow me to see him, so we usually talked over the fence where I was staying.

After some time, Paul advised, "Why can't you just talk to him and just try to see?"

So I tried to see, and from the beginning he was fine. Relationship-wise he was ok, but his problem was financial management. He was also married, which he didn't disclose to me. Instead, he told me, "I was married but I am divorced." He let me know about his kids. We used to visit them in Zimbabwe. And I didn't know he was communicating with the wife all the time, although they were not living together.

. . .

Then I said, "Papers or no papers, if you want to be with me, divorce and marry me. If you want to be with your wife, then go out of my life."

He kept on saying, "I will, I will, I will."

One day he told me, all of a sudden, "I am quitting my job."

I said, "Why quit your job? That job is so secure. Why do you want to leave the job?"

What he said did not make sense to me. But he didn't listen. He went ahead and put in a resignation letter and resigned and they gave him a package of twenty three thousand Pula. The cheque came to me and we went and cashed it together.

I told him, "You keep this money because houses are cheap in Zimbabwe. You can buy a house for five thousand and then with the rest of the money you can do business, since you don't want to work."

So I left him with the money, but it didn't take long before the money was finished. He didn't buy the house when he went home to Zimbabwe. All he returned with was a van of mangoes.

I asked him, "You bought mangoes for twenty three thousand Pula?"

He was tongue tied. "I don't know what happened to the money," he said, expecting me to believe him.

I let it go.

He kept saying we were still together, but there was nothing in his behaviour to show it. I would go to an auction and buy things, and he would take the things and sell them. I told him I didn't like the lifestyle where he sells house things. "It's not my way."

He did little to change.

. . .

The next day in the morning he said he was going to Zimbabwe. He went to Zimbabwe with the ten thousand Pula. He didn't give me a Thebe.

He was in Zimbabwe for a full month and when he came back, the ten thousand was finished. I told him I was moving out. Everything we had in the house—two fridges, beds, wardrobes, a stove, and you name it. I said, "Ok, I don't mind, I will give you all those things. I will start life afresh." And I just took my clothes and my shoes, and left.

"Good riddance," interrupted Dr. Nanny.

It's not finished. My O and A Level certificates, he took them and threw them away and I didn't realise it until much later.

I went to the house where I was going to stay on my own, feeling bitter but relieved. After a week, one girl from near where Noway lived saw me and said, "Eh Miss, ah how come you throw your certificates away? I saw them in pieces."

I couldn't believe it. Certificates are not things to handle carelessly—even mad people know that.

The girl said Noway was seen throwing my certificates, and some kids took the plastic paper and were playing with it, and she only saw them after they had been torn.

I had taken time to laminate my certificates as the best way of protecting them, having grown up where it was all too common for one to lose years of hard earned qualifications to rats and white ants. I went there and everything was in pieces, all gone. No problem, I told myself, there is nothing I can do about it. I phoned Noway and said, "Noway, you decided to destroy my certificates. Why?" "To hell with you," was what I got in reply, and he hung up on me.

. . .

And that is the end of my story with Noway.

There is no doubt he wanted me, but he was somebody who suffered from indecision. He didn't know how to put things together. He was a genuine person, though his financial management was not good. He was somebody who can really assist you well, but I couldn't forgive him for hiding from me the fact that he was married, and for not deciding whether he wanted a future with me or with his wife.

I told myself, "I am not going to have another man. I am not going to have a relationship because it is too depressing."

. . .

My money was disappearing. My will was weakened. Yet I grew in determination.

Chapter 38: Unbearable Comforts of Love

One day, I realised I slept but without sleeping. All my heart was about this guy. But it wasn't at all natural infatuation. No, it was not like that.

. . .

I still have the message on my cell phone, and it still gives me goose pimples when I read it. I store every SMS he sends me. I value my cell phone. I appreciate it, I love it, and I want to have it nearby at all times. I don't want to stay for one hour away from my cell phone. Even in church where we are forced to switch off, I will put my phone on vibration mode and place it somewhere sensitive enough to feel it.

My cell phone is my greatest companion, but it is also my greatest terror. The pain, the bad words, they come through my cell phone. When somebody feels like saying something and he can't face me, he will say it through the cell phone. It has made me experience too much abuse. Without my cell phone, I think I would have suffered less. All the messages, all those things he has been telling me, they are there in that place, in that phone.

When he said that, I told myself, "This is where my life is going to end. If I take this thing hot, hot, I'm a dead person. I will go to Mimboland as a corpse."

I said I must change my attitude. So I changed to save my life.

His anger was very abnormal.

I don't think he can stay with a woman, and I remember there was a day the sister said, "Immaculate, I don't know how your relationship with my brother is, but I have come to realise you are the bravest woman I have ever seen, because even we cannot stay with him."

Even in their house he stays in the room most of the time. If you see him going inside the mother's house, he is going to bathe or to take food from the kitchen.

There are times he will come to my house and tell me, "The day you misbehave you are a finished."

I went to talk to Evodia Skatta. I told her what I was going through, and how I needed to protect myself. "I don't want to die here. I don't have a boyfriend. I don't eat anybody's money. I don't see somebody's husband. But this is what I am going through." Evodia Skatta said, "What?"

As long as he kept using whatever he was using to charm me, he got my money if he wanted it. I never refused. If I didn't have any, I looked for it. I could even borrow and give to him. He was always saying he was borrowing the money, but he would never pay me back.

He doesn't take me around in his car, but when he took the car to the garage and it cost two thousand and thirty nine Pula, he took the money from me, and I don't think I'll ever get it refunded. He drives from work to check on me and see whether I am working with a person, but he never picks me up after work. He is there only to make sure I am not with somebody.

He uses that charm when he wants something from me. If he wants something with me, he can use it the whole night before he comes the next day to ask for it. And I won't refuse. He owes me nearly six thousand Pula.

. . .

When he came the first time to have sex with me, I said, "I want a condom. If you don't use a condom, I'm going to make shout and make so much noise. We are really going to fight."

I think he used that condom because he was scared of the mother. He doesn't come to me when the sisters or the mother can see him. He is so scared of the mother. If it wasn't the fear of his mother, he could have raped me and slept with me without a condom.

If I haven't moved from where I am staying, it is because I know that if I go and stay on my own, he is going to come there and do anything. Even now, I will be sleeping and I won't know that this guy is coming. He has never come to my house me knowing. If he says I want to come there he is already at the door, making me startle.

. . .

After we have sex, he can go for three months without talking to me. If we meet, if I meet him face to face, he will just say, "Dumela." Now I understand what it means to say Botswana men can use women. How can he sleep with me yet treat me like shit?

Not once would he give me a lift in his car. He wouldn't even buy me Fanta or give me water to drink, yet he borrows all my money and refuses to pay back, and gets drunk on Chibuku every day.

I have never eaten anything from him for almost two years. The only thing he sent to me was one day at work when he most surprisingly sent me units for P100, only to turn around and say it was a mistake, and that I should pay him back.

Botswana men can really make a woman feel cheap.

When the mother discovered that Philip was interested in me, she started with her own medicine, but I didn't know.

. . .

I decided to talk to Angel, the generous friend of mine we used to visit, whom you interviewed at length.

Angel was pleased to see me after so many years. I had lots of explaining to do, about why I had kept away for so many years. How I could have been in Gaborone all this while and not passed by to say hello. I was a wicked person, she said, half jokingly, refusing to accept what she termed my "flimsy excuse" that I had dumped my problems enough on her doorstep.

When I accepted my mistake and apologised, she opened the door of her generous heart to the problem that had brought me back to her.

I told her everything, from A to Z, from Noway to Philip. "I don't know what to do," I concluded.

She said she didn't know anyone who could help. She used to know a Sangoma from Malawi, but that was years ago, and the man had since moved on.

I had come full circle, to be contemplating visiting the very Sangomas I used to reject when I first came to Botswana.

Even without a Sangoma, Angel was a great help. Listening and doing her best to console me was soothing. I could buy units for a hundred Pula for my cell phone, and I would talk to Angel and cry until the money finished. Every day I spoke with her, but I didn't share my troubles with any Mimbolander, apart from Evodia Skatta and Paul Mufon. Nobody else knew what I was going through.

Every morning I was crying, daytime I am crying, even my workers at times would say, "Are you sick? What is wrong?"

I said, "No, I'm just missing home."

How did I come to know for sure they were using medicine on me? Although Angel said she couldn't help me when I took my story to her, she couldn't bear me crying the way I did when I was on the phone. So she remembered this other lady she had heard about from someone at work. Upon finding out, she got this woman's number.

The lady was an herbalist. Angel explained to her what I was going through, and the woman said she would check and get back to Angel with her findings.

After a couple of days, the woman came back to her. "I can see this guy's mother doesn't want that girl, but the guy wants her bad," she told Angel. 'He is using medicine to make sure she doesn't go to another man. Why I don't know. I can see the guy is not really a serious person, so why he is doing this, I can't say.

"But unfortunately I can't help. I am just an herbalist. I can't help." She can see, but she doesn't throw bones. She can dream something, but she can't cure beyond her herbs. She said, "I won't lie to you. I can't help that girl. Her problem is a big problem.". . .

One day Angel called me to share something positive. The woman herbalist had called to say a patient had come her way who needed a serious traditional doctor and was ready to pay for her to invite somebody from Mozambique. The man had come and was treating that person and others. If Angel liked, she could ask her friend to come and meet this powerful traditional doctor who could heal what was beyond herbs. So Angel called to let me know.

I went to see this doctor, who asked for the details of the guy using this thing on me. I wrote down the names, phone number and residential area of Philip, everything, and described his build and workplace and handed all this to the doctor, who asked me to go and wait for him.

As luck would have it, word went around Gaborone about this important medicine man in town, and people flooded there seeking his cure, magic and blessings. Amongst them, Philip. When he came, the medicine man was able to identify him from the details he gave of himself.

He was coming to fortify his grip on me.

The doctor asked him to go and bring his mum.

The mother came.

The doctor consulted his bones, looked at them and said, "I can see you are bewitching a foreigner in your yard. Why are you sending your daughters to be throwing poisonous things in her house? They go there and pretend they love her but they are eating with her only to eat her up. Why are you bewitching her?"

The mother said, "I didn't want to kill her. I was just trying to separate her from my son."

The doctor said, "Why are you trying to separate her? If your son doesn't want the girl he will leave her, but your son too is using something on her.

The same person you are using things on to get her away, your son is using things to keep a hold on her, and she can't do anything. She can't go to any other person. Your son makes that girl sit there the whole day, even weekends. She can't go anywhere. She goes to work and comes back or to the shops, but nothing more. Is that the life a human being should live? Why are you being wicked to this foreign girl? What has she done to deserve a fate worse than death?"

Then the doctor tried to talk to Philip to stop using this thing he was using on me. "Why are you using this thing? You are not ready for marriage the way we see you. You are not even a friendly man. You are not a jovial person. You don't want to sit with that girl your mother doesn't want, why can't you leave her to go her way in peace?"

That, my sister, is what I have been going through. I have never had a good life. I don't know the cause of this curse.

Maybe it's my weakness that when I am with you I feel I should treat you the way I want myself to be treated. I don't know whether it is because of that, that men tend to take advantage of me or what.

Angel and the herbalist have been a great support. Last week I told them I would love to go home for a while to renew communion with family, friends and the land of my birth.

I told them I needed a breath of fresh air, a sign of life from this strange and stifling condition of living like a dead girl walking.

They encouraged me and prayed for the means to come my way to make this journey possible. "Greet your parents and eat lots of herbs from the tropical rainforest," the herbalist told me, smiling her satisfaction with knowledge of the charming natural environment of my country. "You'll need all the energy you can muster to overcome the forces that hinder the good life for you."

Now that you've surprised me so delightfully when I least expected, I am going right after this to book a flight and buy a ticket to Mimboland.

After thirteen years of a life tortured by worries, I want to rediscover what it means to socialise without having to look behind my back. I want to be able to talk freely and feel like a human being again. At 33, I feel the joys of womanhood passing me by.

Ordinarily, I wouldn't be wishing this, given the brutal pride of power gone wild back home, but my traumas here have drained me. I have gone through too much. As my mother would say, if you see a rat running towards fire, know it is being chased by something even more terrifying.

I need time to regain my dignity, even if it means my hands and legs are going to be broken by the blows of excited rifles and batons. I need family, friends. I need people and places I knew. I need to reconnect to feel human again.

Conclusion

Migration is a complex and multifaceted phenomenon involving various dimensions of human mobility in claiming and negotiating inclusion and belonging. It challenges rigid and bounded distinctions between insiders and outsiders in favor of a more flexible understanding of belonging in tune with the frontier reality of Africans as bridging intimacy and distance. Immaculate's story highlights the thrills and challenges of forging relationships in host countries, host communities, and beyond in the pursuit of success and self-fulfillment. The story addresses the perplexing question of what it means to be a person of African descent living as a stranger in another African country. Life away from a place called home is informed by memories of home and by social networks and relationships at home and elsewhere. While a single perspective is neither desired nor sufficient, Immaculate's story emphasizes the significance of sociality and relationships in how being and belonging are translated from abstract claims into everyday practice. Mobility, connections, and interconnections are emotional, relational, and social phenomena best understood as complex, contradictory, and messy realities that defy prescriptiveness, predictability, insensitivities, and caricature. Through Immaculate, we are introduced to the predicaments of being mobile but also to the inadequacy of the legal mechanisms elaborated and employed by states to regulate inclusion and exclusion in a world of accelerated and flexible mobility. Hers is a complex story that calls for complex approaches.

Immaculate and her story invite us to focus on the lived experiences and web of relationships that shape and are shaped by intra-African migrants in and beyond their host states and communities. Personal and collective success is critical to migrants, as are their social networks and cultures of interdependence and conviviality. Choice and chance are good bedfellows in the construction and management of social networks and relationships (Owen 2011). This is hardly surprising in a world where agency and contingency are like Siamese twins. By choice or chance, Immaculate successfully draws on her relationships with others and on the cultures of interdependence and conviviality that have shaped her from childhood, to maximize her opportunities as a young, mobile African woman. Her story takes us through the relationships she forges in her efforts to navigate, negotiate, and contest various constraints and practices of belonging imposed by the logics, histories, and politics of hierarchies, dichotomies, boundaries, and exclusions. The story stresses the need for conceptual flexibility and empirical substantiation. It equally challenges social scientists to look beyond academic sources for ethnographic studies or accounts of how such flexible and nuanced understanding of mobility and interconnections in Africa play out in different communities, states, and regions of the continent.

Hierarchies in Immaculate's world are not that dissimilar to hierarchies in the world of knowledge. To what extent, if at all, is science superior to literature? When is literature entertainment? And when is it an insight into human nature? To what extent could science double as entertainment? These questions have preoccupied many. Eggington (2011), for example, argues that fiction plays "a profound role in creating the very idea of reality" that science seeks to explain. He cites the example of Cervantes, in *Don Quixote*, who "crystallized in prose a confluence of changes in how people in early modern Europe understood themselves and the world around them."

In the twenty-first century, ethnographic fiction or "anthropological novels" may have a role in revealing the "fiction" behind multiple forms of discourses of dominance. While the idea of the ethnographic novel is gradually gaining suffrage in anthropological circles (Gupta & Ferguson 1997), Shule (2011) believes that works such as *Intimate Strangers* suggest "new possibilities in African literature." This augurs well for the disciplinary, conceptual, and methodological intimacies and flexibilities suggested in this chapter.

Note

1. The term *makwerekwere* is generally used in a derogatory manner to refer to African immigrants from countries suffering an economic downturn. Stereotypically, the more dark-skinned a local is, the more likely he/she is to pass for a *makwerekwere*, especially if he/she speaks Setswana. BaKalanga, who tend to be more dark-skinned, are also more at risk of being labeled *makwerekwere*. In general, the le-/ma- (sing./pl.) prefix in Setswana designates someone as foreign, different, or from outside the community. It is not used just for ethnic groups but for any group or profession that seems to be a bit different from the average.

References

Adichie, C. (2009) "The Danger of a Single Story," TED Conferences, www.ted.com/talks/lang/eng/chimamanda_adichie_the_danger_of_a_single_story.html, accessed September 29, 2011.

Bakewell, O. (2008) "In Search of the Diasporas within Africa," *African Diaspora* 1(1–2): 1–27.

Eggington, W. (2011) "*Quixote*," *Colbert and the Reality of Fiction*, http://opinionator.blogs.nytimes.com/2011/09/25/quixote-colbert-and-the-reality-of-fiction, accessed September 29, 2011.

Geschiere, P. (2009) *The Perils of Belonging: Autochthony, Citizenship, and Exclusion in Africa and Europe.* Chicago: University of Chicago Press.

Giblin, J. (1999) *Issues in African History*, www.uiowa.edu/~ africart/toc/history/giblinhistory.html, accessed September 29, 2011.

Gupta, A. & J. Ferguson (1992) "Beyond 'Culture:' Space, Identity, and the Politics of Difference," *Cultural Anthropology* 7(1): 6–23.

Gupta, A. & J. Ferguson (1997) "Discipline and Practice: 'The Field' as Site, Method, and Location in Anthropology," in: A. Gupta & J. Ferguson (eds.), *Anthropological Locations: Boundaries and Grounds of a Field Science*. Berkeley: University of California Press, pp. 1–46.

Kopytoff, I. (1987) "The Internal African Frontier: The Making of African Political Culture," in: I. Kopytoff (ed.), *The African Frontier: The Reproduction of Traditional African Societies*. Bloomington: Indiana University Press, pp. 3–84.

Landau, L. B. (2011) *Exorcising the Demons Within: Xenophobia, Violence and Statecraft in Contemporary South Africa*. Johannesburg: Wits University Press.

Neocosmos, M. (2010) *From "Foreign Natives" to "Native Foreigners:" Explaining Xenophobia in Post-apartheid South Africa: Citizenship and Nationalism, Identity and Politics*. Dakar: CODESRIA.

Nyamnjoh, F. B. (2005) "Fishing in Troubled Waters: *Disquettes* and *Thiofs* in Dakar," *Africa* 75(3): 295–324.

Nyamnjoh, F. B. (2006) *Insiders and Outsiders: Citizenship and Xenophobia in Contemporary Southern Africa*. London: Zed/CODESRIA.

Nyamnjoh, F. B. (2009) *Married But Available*. Bamenda: Langaa.

Nyamnjoh, F. B. (2010) *Intimate Strangers*, Bamenda: Langaa.

Nyamnjoh, F. B. (2011) "Cameroonian Bushfalling: Negotiation of Identity and Belonging in Fiction and Ethnography," *American Ethnologist* 38(4): 701–713.

Owen, N. J. (2011) " 'On se Débrouille:' Congolese Migrants' Search for Survival and Success in Muizenburg, Cape Town," Ph.D. Thesis, Department of Social Anthropology, Rhodes University, Grahamstown, South Africa.

Peberdy, S. (2009) *Selecting Immigrants: National Identity and South Africa's Immigration Policies 1910–2008*. Johannesburg: Wits University Press.

Shule, V. (2011) "Intimate Strangers," *English Academy Review* 28(1): 121–123.

Stolcke, V. (1995) "Talking Culture: New Boundaries, New Rhetorics of Exclusion in Europe," *Current Anthropology* 36(1): 1–24.

Wolfe, P. (1999) *Settler Colonialism and the Transformation of Anthropology: The Politics and Poetics of an Ethnographic Event*. London: Cassell.

Zeleza, P. (2011) *In Search of African Diasporas: Testimonies and Encounters*. Durham: Carolina Academic Press.

Index